World University Library

The World University Library is an international series of books, each of which has been specially commissioned. The authors are leading scientists and scholars from all over the world who, in an age of increasing specialization, see the need for a broad, up-to-date presentation of their subject. The aim is to provide authoritative introductory books for university students which will be of interest also to the general reader. Publication of the series takes place in Britain, France, Germany, Holland, Italy, Spain, Sweden and the United States.

Gunnar Thorson

Life in the Sea

Translated from the Danish by
Manon C. Meilgaard and Alec Laurie

World University Library

McGraw-Hill Book Company
New York Toronto

Reprinted 1972
Library of Congress Catalog Card No: 73–118405

Photoset by BAS Printers Limited, Wallop, Hampshire, England
Manufactured by LIBREX, Italy

Contents

PART TWO

Introduction

The sea covers nearly 71 per cent of the earth's surface, but at present it contributes less than 2 per cent of our food supply. Since the world is steadily becoming more and more overpopulated, and since the number of people who are undernourished, if not starving, is steadily increasing, the exploration and exploitation of the sea is plainly a vital matter.

More than 90 per cent of all fishing takes place in coastal waters where the depth is less than two hundred metres, and the area of those waters is no more than 7 per cent of the world total; 88 per cent of the total sea area has a depth of over one thousand metres, and it is the potentiality of these deep waters which has hardly been touched so far. If we are to exploit the deep ocean we have plainly a very considerable task ahead of us.

This is not to suggest that the exploration of the great ocean depths has been neglected. A good deal of work has been done, but only in the last hundred years. The reason for this is that until 1858 it was commonly believed that living organisms hardly existed below a depth of 500–600 metres, and as a corollary to this it was assumed that approximately 90 per cent of the sea bed was completely devoid of life. A change in attitude came about in the 1860s when one of the first submarine telegraph cables was brought up from a depth of over 2,000 metres in the Mediterranean and was found to be covered with living animals which were obviously bottom dwellers. This discovery opened up deep-sea exploration, and many expeditions were organised, the most notable being the British *Challenger* Expedition of 1872–5.

With the passage of time the techniques of deep-sea work were enormously improved until, in 1951, the Danish *Galathea* Expedition dredged up living animals from a depth of 10,150 metres from the Philippine Trench. This proved that life could exist at even the greatest known depths of the oceans. Nine years later (and only a hundred years after the broken telegraph cable incident), Jacques

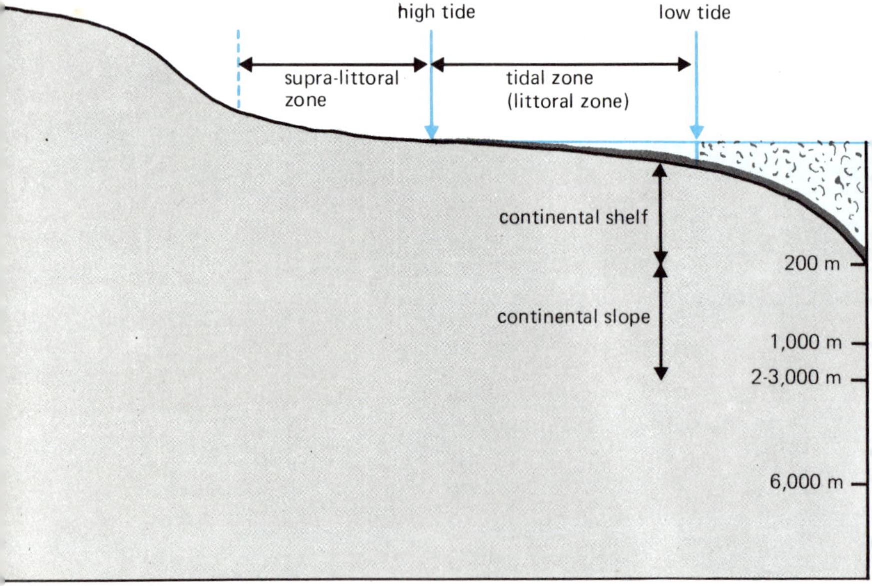

Piccard and Donald Walsh successfully descended to a depth of 10,910 metres in a specially constructed vessel called a Bathyscaphe. This descent was in itself a superlative achievement, and was especially significant in that it opened up an era of exploration in which people could actually see and study marine life at great depths.

A recent list shows that there are more than forty research submersibles of many different types, constructed for work at different depths. The United States, France and the Soviet Union are the leaders in this field. One of them, *Alvin*, built by the American navy, seems to be particularly efficient for biological work. It carries one pilot and one or two observers and, accompanied by its mother ship, it has already successfully completed hundreds of dives, some of them to more than two thousand metres, and lasting as much as eight and a half hours. *Alvin* was obviously well built; on one occasion she was attacked head-on by a large sword fish weighing about one hundred kilos. The sword fish broke its bill and died, but *Alvin* was unscathed. It goes without saying that the extremely hard alloys which are now being developed for aeronautics will be equally useful

The diagram explains the terms commonly used to refer to areas of the sea and shore.

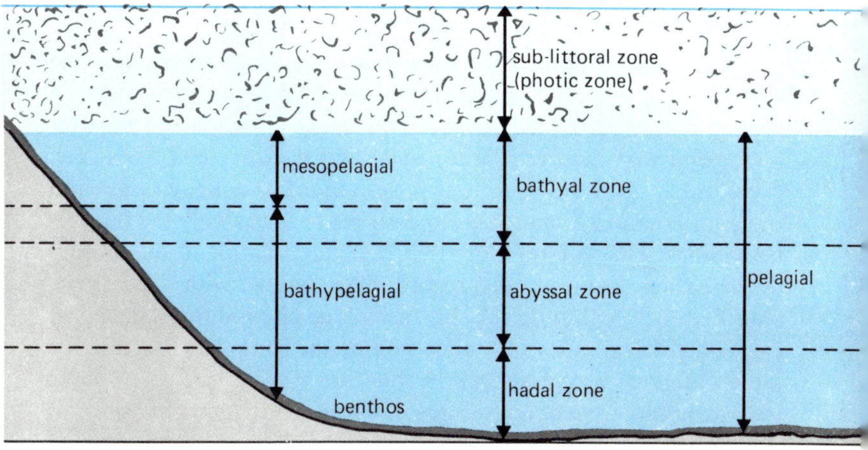

in the development of submersibles for use at great depths both now and in the future.

If we consider that the *Challenger* did all her deep-sea soundings laboriously by manpower and using a very long grass rope, while the *Alvin* slips quietly through the water propelled by the most sophisticated gear and equipped with the most modern instruments, we have some sense of the progress of practical oceanography over the last century. But this progress means very little unless we realise that the contrast between the work of these two ships is qualitative, and not quantitative. The fact still remains that there have been relatively few expeditions, and that all they have been able to do is to sample the oceans in outline, so to speak.

Most of the land masses, including the most desolate wastes and polar regions have been well served by explorers, by bases and research stations, and from these detailed studies of the soil, and the animal and plant life have been made. But the investigation of the oceans has, until recently, been based on the work of quite a modest number of ships, and the samples and specimens procured have been gathered during the performance of a predetermined programme which, when seen on a chart, very often resembles no more than a zig-zag course across the oceans. This is not to minimise the extremely good work and excellent results that have accrued from such expeditions. Nevertheless, we must regard the knowledge and material obtained from these expeditions so far as only the humble beginnings. They are no more than a slender sample of the totality of the oceans. We must also regard these samples with some suspicion because the biological and hydrological conditions that may exist only a few miles on either side of the sampling line may be quite different; we cannot take a thin line of observations as being truly representative of a vast area of ocean. One can emphasise this point by saying that the combined efforts of all deep-sea research up to the present cover less than one twenty-five millionth of the total bottom area of the deep seas.

From its beginnings in the latter part of the nineteenth century up to the 1950s, deep-sea research has been at best a spasmodic affair. Research vessels are costly to run, and one finds that most of these expeditions have been sponsored either by wealthy foundations or, less frequently, by governments in times of prosperity. Expeditions sponsored by wealthy foundations have been almost entirely academic, and nonetheless valuable for that; government-sponsored expeditions have been equally academic, although at times there have been some practical reasons for a particular series of investigations, as for instance the *Discovery* investigations, which had as their raison d'être an enquiry into the stock of southern whales and everything that was relevant to the maintenance of that stock. Furthermore, it would be no exaggeration to say that government expeditions have been carried out from time to time as a method of showing the flag and as a form of international prestige. This is not to denigrate the excellent results that have been achieved, but it is fair to say that these expeditions have been very similar to the exertions of small bands of explorers setting off into the unknown, recording what they find.

Times have changed, however, and the 'developed countries' have become deeply aware of the population explosion, which has outstripped estimates to such an extent that the prognosis is now alarming. It has therefore become a matter of extreme urgency to attack the problem of what can be provided from the sea simply in order to feed this expanding population. And so, in the 1960s, a new era of intensive exploration dawned. Its terms of reference were well set out by President Kennedy in a message to Congress in March 1961:

> The seas offer a wealth of nutritional resources. They already are a principal source of protein. They can provide many times the current food supply, if we but learn how to garner and husband this self-renewing larder. To meet the vast needs of an expanding population the bounty of the sea must be made more available. Within two decades our own nation will require over a million more tons of sea food than we now harvest.

From now on the principal motivation for intensive research will be the hunt for food and yet more food; but as yet we have no certainty from what part of the sea, or from what depths in the seas, that food will come.

The shallow-water areas, from which comes most of the present protein harvest and which are fairly well known already will be investigated using modern equipment in the hope that we may learn more about the habits, life history and general behaviour of marine animals in those waters. Thorough exploration of the really deep oceans will also be intensively carried out, not because there is any immediate indication of a usable food source, but simply because there is no telling, in the present state of knowledge, what will turn up. As we shall see later in this book, the deeper you go the less life there is in the sea, and we can expect that if the deep oceans produce a substantial crop of protein, that crop will be found in the top few hundred metres. But that in itself is no reason for not exploring the utmost depths when a fully equipped research ship can investigate both the upper layers and the abyss at the same time. What this all amounts to is that the research of the future will still be very largely academic, but it has to be done because at present we simply do not know enough about life in the oceans.

We have stressed the importance of an all-out attack on the oceans, but it is not to be thought that this is something for the future. Already a great deal has been done in the last few years; for instance the Russians now have a very large expedition ship of 5,000 tons working the year round in both the Atlantic and Pacific Oceans. This ship carries sixty scientists as well as a normal complement of navigation officers, deck-hands and engineers. The Americans are preparing a network of permanent research ships which will be stationed at strategic points in the major oceans. Since 1965 there have been reports of forty American vessels busy on various research projects, and the government budget for marine research for 1966 was fixed at 141 million dollars. One of the first onslaughts on a specific problem was carried out between 1959 and 1965 in the Indian Ocean, a hitherto neglected region. The object was to assess the yield of food that might be available, which, if present, would be of vital importance to the huge human populations living around its coasts. This was a genuinely combined operation, international in character. A fleet of research vessels from thirteen nations manned by scientists from twenty-three countries took part. Already, after only three years, more than four thousand pages of results have been published, and there is much more to come.

Since this book will be mainly concerned with the life and habits of marine organisms and their relations to each other and their environment (ecology), it will be convenient to mention here some of the techniques which are being used on a smaller scale in oceanographical research work.

We have mentioned the Bathyscaphe and other submersibles, which represent the direct approach to submarine life. Supplementing these, and far less costly to build and operate, are the implements of indirect approach. Into this class falls, for example, a French invention of a deep-sea sledge, fitted with spotlights and a closed circuit television system. As the sledge is towed along the seabed, animals suddenly appear on the receivers' screen, are seen clearly for an instant, and then disappear again, rather in the way that a road rolls backwards in the headlights of a car.

Britain too has pioneered underwater television, in a slightly different form. The complex of spotlights and television cameras is suspended just above the sea bed, and as the research ship slowly drifts along, a running presentation of animal life appears on the screen in the ship's laboratory. This technique has been refined to the point where it is now possible not only to identify with some certainty the different species of bottom animals appearing on the screen but also to calculate their density per square metre. Another variant consists of attaching the apparatus to a mid-water trawl in order to observe both the behaviour of the trawl itself and also the escape reactions of the fishes. This is an example of a device which may well serve a dual purpose, in that it is useful both for basic research and also for improving the design of trawling gear, so as to increase the catch.

There is no doubt that these examples of closed circuit television will be very useful, but for pure research they have one disadvantage: fish are extremely sensitive to shock-waves generated either by other fish or by any moving object. If therefore we are to study the *natural* behaviour of fish and other marine animals we need an apparatus that can act as a silent and motionless interloper. Just such a device has been installed on the sea bed of the Bahamas. It consists of a permanent television camera and microphone, and is connected to the shore by a multicore cable which transmits both vision and sound signals to a shore laboratory more than four kilometres distant. By remote control the operator can switch on and off, and also rotate the camera in any direction. It may be asked what the purpose of the microphone is. The answer is that both the development of electronic gear and the observation of frogmen have revealed that below the surface there is a constant hubbub of grunts, squeaks and growls emanating from a great variety of animals. These noises are worth investigating; indeed anything that can add to our knowledge must not be overlooked. The activities of frogmen are by now familiar to most of us through films created by Jacques-Yves Cousteau, the great pioneer of free diving (that is, without an air-line to the surface), and by Hans Hass. Frogmen, whether they are free swimming or saving energy (and hence oxygen) by holding on to an electrically propelled scooter, are extremely useful in shallow waters, say down to seventy metres. But there are two physiological dangers which limit their activity. The first is that a prolonged sojourn, using the normal oxygen-nitrogen mixture of air, can be dangerous. Below seventy metres nitrogen dissolving in the blood affects the brain, producing a condition of irresponsibility and carelessness known medically as 'nitrogen narcosis'. Cousteau's own definition ('*ivresse des profondeurs*', or 'drunkenness of the deeps') is more descriptive. Because of the toxic effect of nitrogen at high pressures there is a limit of about ninety metres to the depth to which experienced divers can go without needing special and costly mixtures of oxygen and helium.

The second hazard is supersaturation of the blood with dissolved nitrogen at any depth below 13·5 metres. So long as the diver stays down this has no effect, but if he surfaces too quickly the decompressed nitrogen comes out of solution, forming bubbles in the joints, and in acute cases in the central nervous system. The effects of the 'bends' or caisson sickness range from mild discomfort to paralysis and even death.

The remedy is simple: there are in existence precise and well tested tables of the time that a returning diver must pause at different depths in order to let the nitrogen diffuse normally through the lungs. If this stage decompression procedure is strictly followed there is no danger, but it is very time consuming.

This leads us directly to the development, in the past five years, of underwater accommodation for divers, known as 'sea-labs'. The idea is to set up submarine living and working space consisting basically of a moored hollow steel chamber through which there is direct access to the sea by way of an opening at its lowest point. The living space is ventilated by air provided at a pressure appropriate to the depth, and the object is to provide divers with a pressurised home for days or even weeks. When the time comes for a spell on shore the divers need undergo only a single stage decompression, instead of going through that tedious procedure perhaps as often as several times a day.

Cousteau pioneered the sea-lab in the Mediterranean. The American prototype was *Sea-Lab I* off Bermuda in 1964, followed by *Sea-Lab II,* southern California, 1965, and *Sea-Lab III,* also off southern California. A fourth installation, *Tektite I,* started operations in sixteen to seventeen metres off the Virgin Islands early in 1969; four scientists work there for three to four months before returning to dry land. The American aquanauts are not content with telephone communication with the mother ship and television for their entertainment; they are also training bottle-nose whales and sea lions to act as postmen for collection and delivery of small articles between the sea-lab and mother ship, and even for assisting in locating missing frogmen. In Britain a tighter budget has resulted in smaller sea-labs designed for a crew of two, and a stay of one week.

Lastly, and on a gigantic scale, there are plans to use orbiting satellites for reporting ocean currents (distinguishable by colour changes), and aggregations of plankton and shoals of fish. Such observations will enable research ships to take the shortest route to such indications instead of laboriously criss-crossing the oceans.

Such are the latest tools of oceanographical research. Behind these we have a vast array of dredges, trawls, plankton nets of varied diameter and mesh, water and temperature sampling and recording gear, and many other devices ranging from sophisticated isotope experiments to the homely spade for digging up specimens at low tide.

This book is basically a report on the present state of our knowledge of marine biology and on the outstanding problems calling for solution. What we have to say can be regarded only as a springboard for the intensive research that our present food situation demands.

PART ONE

1 The sea as habitat

The most recent research shows that life has existed on earth for not less than 3,500 million years, and it is now generally accepted that the largest and deepest oceans have existed for as long a time. There are indications, too, that the first living organisms originated in the sea. Here, the first creatures could live freely, without competitors, enemies or parasites – sole inhabitants of the vast ocean. It is also known that the oldest fossils are of sea and not of land creatures, so that there must have been a very long period in which marine fauna and flora were able to develop—not less than 600 million years. By contrast, the first land animals appeared only 300–350 million years ago. We have already said that the sea covers more than two-thirds of the earth's surface, but then we were talking in terms of comparative *areas*. If we compare the *volumes* of living space on sea and land, we arrive at a very different figure. Most land organisms live in a vertical zone extending from the tree tops to about one metre below the soil surface – that is, about twenty metres. (Also, in calculating the living space of land organisms we must deduct the polar wastes and large deserts in which life is either scarce or absent.) The sea, on the other hand, is fully three-dimensional. The average depth of the seas is around 3,800 metres, and since even the greatest depths are inhabited we can calculate that this 1,400 million cubic kilometres of living space is roughly three hundred times larger than that available for life on land.

Salts

All marine organisms have one thing in common: by definition they live in salt water, but the degree of salinity varies from place to place, and is usually expressed as 'parts per thousand' (‰) and not as a percentage. In the open sea the salinity varies from around 34‰ in polar waters, where melting ice adds fresh water in summer, to 37‰ in tropical waters, where evaporation from the surface is high.

1·1 A diagram to illustrate the role of plankton. Fish in both the free-water masses and on the bottom are directly or indirectly dependent on this source of food. (After G. Thorson 1952.)

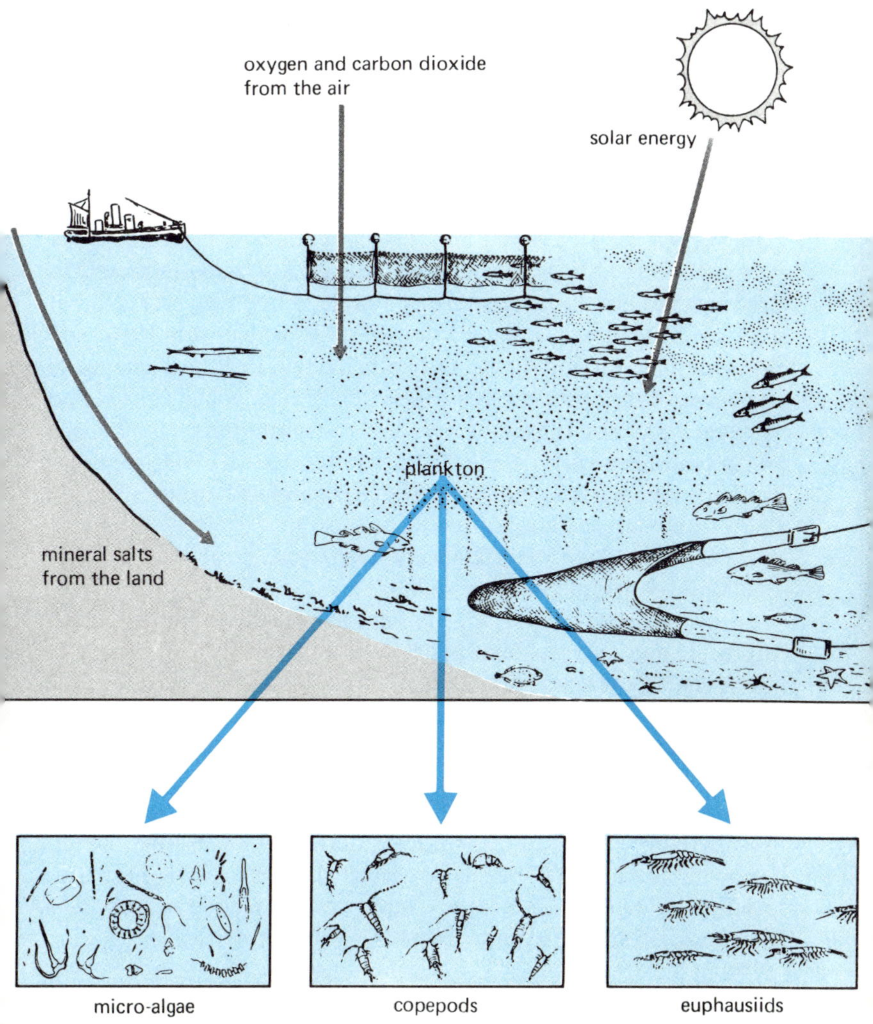

The Red Sea, with up to 43‰, has the highest salinity of any open-sea area, while the Baltic has the lowest, at about 7‰.

About 78 per cent of the sea's salt consists of sodium chloride; but there are also important amounts of magnesium chloride, magnesium sulphate, calcium sulphate, calcium carbonate, and potassium sulphate. There are also many other salts in such small amounts that they can be measured only in parts per million or parts per thousand million.

Some of the more scarce salts were originally found indirectly in 1865 by the Danish scientist Georg Forchammer, who realised that a chemical analysis of these salts, existing in relatively high concentrations in marine plants and animals, revealed the existence in sea water of elements such as copper, lead, zinc, cobalt, nickel and barium, which are present in very low concentrations. None of the salts could have been detected in sea water itself by contemporary methods of analysis. Today, using more refined techniques, we know that, for instance, brittle stars contain concentrations of silver and zinc which are 100,000 and 1,000 times greater respectively than those of the surrounding waters.

In the sea, as on land, all life depends on plant matter, which captures solar energy. Plant matter in turn depends on a supply of the so-called nutrient salts, especially phosphates and nitrates. The absence of even one of these compounds, which are usually found in rather small quantities, is enough to stop plant production. The growth of marine algae, like any other plant growth, can take place only where there is sufficient light – that is, in the 'photic zone' – and it is therefore limited to the surface waters. This means that the top layers of the sea tend to be quickly depleted of nutrients. Continuous production depends on a steady supply of nutrients transported to the surface from below, and this in turn requires a vertical circulation of water masses.

Some of the richest production is found in places where a warm and a cold current meet. The waters of two such currents have, usually, different specific gravities, and this leads to a vertical circulation which brings fresh nutrients to the surface. Oceanographers say that in such conditions the sea 'ploughs itself', a ploughing which may go as deep as four thousand metres. This kind of mixing is found, for example, on the Newfoundland Banks, off north-west Iceland, and in the Lofoten area – all important fishing grounds.

Sometimes two surface currents each produce a poor crop of algae; but when they meet and mix the result is a huge increase in production. This is almost certainly because one of the two currents has lacked one of the salts vital for algal growth, while the other has been carrying a surplus of that very salt. When the water masses mingle they are both able to resume full production.

Salinity and living matter From a purely physical point of view salinity is important to plants and animals. The greater the salt content, the greater the buoyancy and the easier it is for marine organisms to float. But a high salinity, from which a marine organism can never escape, presents a physiological problem, especially in advanced forms of life, in that there is a difference between the salinity of the ambient sea and that of the body fluids. This raises the question of *osmosis* – the tendency of water to move across a semi-permeable membrane from a zone of low concentration of salt to a zone of high concentration.

In fresh-water fish the salt concentration in the blood and body fluids is much higher than that of the surrounding water. A large proportion of fresh water is, therefore, carried into the body of the fish via the gills, and while this must of course be excreted, the necessary salt concentration must be retained. Hence, the kidney of a fresh-water fish has the function of maintaining the salt level in the body, while excreting large amounts of urine with the lowest possible salt content. But with sea-water fish the situation is reversed.

The salt content of the blood and body fluids is only 30–50 per cent of the sea's salinity, so that together with the food transported through the intestine, the fish receives a greater amount of salt than it needs. Most sea fish solve this problem by excreting very small quantities of urine containing a very high concentration of salt. This means, in effect, that the kidneys have to work against the normal osmotic trend and in the process this inevitably calls for an expenditure of energy, which we call 'active transport'.

In primitive fish such as sharks and rays, the salt content of blood and sea water is about the same, and this is true also of many of the lower invertebrate forms – sponges, jellyfish, sea anemones, corals and echinoderms – in which the gastral cavity and even a primitive circulatory system are in direct connection with the sea water.

Salinity barriers Many sea creatures, and especially those that live in the relatively constant environment of the oceans, are very sensitive to any change in salinity. A decrease from the average 35‰ of the oceans to the 27–30‰ often found in shallow-water coastal areas, is enough to exclude many oceanic forms. In the case of echinoderms, amphipods and isopods it has been shown that 75 per cent of the species found in the Skagerrak at a salinity of 34‰ no longer exist at the entrance to the Baltic where the salinity has dropped to 30‰. On the other hand, a number of forms in coastal areas can tolerate a salinity as low as 15‰, and a small group of special brackish-water species is able to tolerate an even lower figure. This last group has an advantage in being able to live in an area unsuitable for most other animals, so that it suffers less competition. Nearly all these brackish-water species are related to true marine forms; but fresh-water fauna seems unable to adapt itself to a salinity of more than a very few parts per thousand.

The opposite extreme is found in certain marine animals, which are able to survive in lagoons and salt marshes where, because of evaporation, salinities are often much higher than those of the open sea. We know of marine species, protozoa, flat worms, polychaete worms and ostracods which can survive and grow at salinities from 100–180‰, while many other marine-animal groups have representatives which can live at 60–80‰. Most of these hypersaline species, however, can only vegetate: they cannot reproduce. This means that their populations can be maintained only by an influx of larval forms from the open sea. Furthermore, survival in hypersaline conditions is critically linked with water temperature; for example, in a hypersaline lagoon in Texas, eighty species could stand a salinity of 50‰ at 20°C, but only thirty-two species survived at the same salinity if the temperature rose to 30°C.

Temperature

As we have said, the salinity of oceans is fairly constant, but temperatures vary greatly. In polar regions there are water masses with surface temperatures of −1·5°C all the year round. At the opposite extreme the average temperature in the Persian Gulf is 24°C, and in the gulf's shallow coastal areas it is often over 30°C. Surface temperatures vary widely with latitude, but their daily and seasonal fluctuations are nowhere near as great as those of land areas in the same latitudes. The principal reason for this is that water has the highest specific heat of any known substance and therefore requires more heat to raise its temperature than any of the substances found on land; conversely, water takes longer to cool down than any other substance. Also, as we shall see, a great deal of the sun's heat fails to penetrate the water at all.

While the temperatures of surface areas fluctuate with the seasons, all water masses below a depth of two hundred metres have roughly the same temperature throughout the year. Just as land temperatures decrease from the tropics to the poles, so the sea temperature decreases vertically from the surface to the great depths. Apart from sea areas such as the Mediterranean and the Red Sea, where special conditions prevail, the temperature at 2,000–3,000 metres never rises above 4°C even at the equator – about the same as that of deep water in polar regions. This similarity is to be expected, since cold polar water is heavier than warm tropical water and therefore sinks below the surface. Cold water from the Antarctic in particular flows northwards deep below the surface layers, and can be traced far up into the northern hemisphere.

The constancy of deep-water temperatures means that many sea creatures live under very stable circumstances, so stable, in fact, that animals from the shallow Arctic sea bed have been found living at depths of 2,000–3,000 metres in equatorial waters. The idea that some species occur in both Arctic and Antarctic shallow areas, thus having a 'bi-polar' distribution, has, however, been abandoned by most marine biologists.

Light

As we have said, life in the sea depends on microscopic algae (the sea's *primary producers*). It is not surprising, therefore, that these algae are to be found close to the surface and within range of daylight. As a general rule, we can say that a light intensity of no more than 1 per cent of noon sunlight is enough for effective photosynthesis, and yet 90 per cent of the world's water masses are in total darkness. The first light barrier is the surface itself. In polar regions, where the sun is never very high, the light always strikes the surface at an oblique angle and a considerable percentage is reflected. Even during the polar summer, when the sun is at its highest and never sets for several weeks, very little light actually penetrates the water. Also, in rough weather, when the surface is choppy, as much as 60 per cent of sunlight is stopped or reflected in the uppermost few metres, the increase being due to the presence of many bubbles on or near the surface, formed by breaking waves.

What we have just said applies to the visible spectrum; other wavelengths invisible to the human eye fail to penetrate much below the surface. Infra-red rays heat the actual surface of the sea but cannot penetrate far. Ultra-violet rays penetrate to a maximum of three metres, and this, incidentally, is just as well, because ultra-violet light is injurious to many planktonic eggs, larvae, and adults, as well as to several species of algae and sedentary animals in shallow water. Ultra-violet rays are also fatal to bacteria, but as they are so quickly absorbed their good or bad effects can hardly be traced more than three metres below the surface. It is possible, however, that ultra-violet light in the surface layers plays a vital role in the formation of several important vitamins in micro-algae.

Now let us follow the fate of such visible light as was not reflected. As a rule the red rays are very quickly absorbed. This would seem to militate against photosynthesis in the red part of the spectrum, red light being particularly effective in this process. In the Sargasso Sea, where the water is more transparent than anywhere else in the oceans, only 1 per cent of the the red light penetrating the surface is visible at 55 metres, while 1 per cent of the yellow-green and violet rays are still visible at 95 metres. Blue rays, which can also be used in photosynthesis, have the greatest penetration, reaching 150 metres before being reduced to 1 per cent. In most other tropical seas blue is reduced to 1 per cent at about 100 metres.

These conditions apply to very clear water, but sea water often contains substances in solution and large quantities of microscopic particles in suspension. These can have a marked and selective effect on the absorption of different colours. In the Atlantic off North Africa, for instance, blue rays seem to be absorbed twice as quickly as yellow-green rays. Coastal waters, on the other hand, are generally found to provide maximum transparency in the yellow-green portion of the spectrum, and daylight is reduced to 1 per cent of its surface value at between 10 and 30 metres' depth.

It is interesting to note that macro-algae (sea weeds) in coastal waters can be adapted to selective colour penetration. Green algae exploit the energy of the red and blue parts of the spectrum and are therefore adjusted to quite shallow water; then come the brown algae, which can thrive without red light and are concentrated between 5 and 15 metres; and lastly the red algae, which reflect red but can use blue and green light, found between 15 and 30 metres.

We have mentioned the transparency of the Sargasso Sea, but this is quite exceptional. Measurements from the continental slopes to deep water in both the Atlantic and Pacific show that the various colours of the spectrum penetrate to only half the depths that they do in the Sargasso. It seems that the depths of even the most transparent ocean areas are in total darkness. Using sensitive instruments, however, a very faint trace of light appears to penetrate considerably further down, and this can be utilised by some organisms. A light intensity of only a fraction of a thousandth of the surface light has been detected at a depth of 800 metres in the Mediterranean, and 950 metres in the Caribbean.

Light and vertical migration Many species of organisms in the upper layers make regular vertical migrations, moving towards the surface at night and away from the surface during the day. It has long been established that these movements are directly influenced by light; but the same kind of migrations have now been found to take place among planktonic organisms as far down as eight hundred metres – the maximum depth at which light can be detected instrumentally. These creatures must be extremely sensitive to light, and recent physiological experiments seem to support this assumption. It is now estimated that a deep-sea fish with large eyes can perceive light about a hundred times weaker than the limit of perception of a human eye. It follows that in very clear ocean water such a fish would be able to detect daylight for a brief period, when the sun is highest, even at nine hundred to a thousand metres, and begin to distinguish objects at a light intensity only a few times brighter. Fish seem to perceive green light better than blue; but many crustaceans respond to blue light better than green. Since bluish-green light penetrates further than pure green, this suggests that some crustaceans can perceive daylight at even greater depths than fish.

A number of fish living at about five hundred metres have huge eyes with an unusual structure. The same is true of some squids, which have eyes with a diameter of up to thirty-seven centimetres. Some of these forms undertake diurnal migrations, like the plankton of the upper layers but in contrast with true deep-sea animals which live in total darkness. The latter do not have exceptionally large eyes; indeed some of them are blind.

Luminiferous organisms The sun, however, is not the only source of light in the sea. There are many organisms that produce their own; these range from unicellular plants to jelly fish, shrimps, squids and fishes. In warm waters, a group of algae, dinoflagellates, light up when disturbed so that the sea appears to be on fire. In deep water, luminiferous organisms can be detected by a light-sensitive underwater camera, which is triggered by the animal's flash. Some fishes carry light organs in special patterns along the body; these patterns vary from species to species, and often from male to female – an excellent aid for individuals of the same species but of different sex to identify each other in spite of the darkness. In other species light organs are used to attract prey, or to frighten or give warning of enemies. For example, certain species of squid emit a phosphorescent cloud instead of the more usual sepia. But in many other cases scientists are still wondering what is the significance of light emission. Most organisms do not emit light continuously; they can switch their lights on and off by nerve impulses. It is interesting and puzzling that many luminiferous species, especially shrimps and fish, do not live in the dark deep waters but only 150–500 metres below the surface, and that the ability to emit light is particularly conspicuous in species which make the diurnal migration.

New and extremely sensitive photometers have recently been lowered to the depths where bioluminescent animals are plentiful, and these have shown that the darkness is constantly interrupted by a large number of flashes. Such flashes are weak and very brief, but no doubt they can be detected by others of the same species and also by nearby members of different species, which may be helped to orientate themselves.

In deep water off New York the frequency of flashes has been measured at about 160 per minute at 100 metres, 90 per minute at 900 metres, and only one per minute at 3,750 metres – the greatest depth at which flashes have so far been registered. Flashes are usually blue or blue-green; in fact they are in just that narrow part of the spectrum to which most deep-sea animals are most sensitive. Some luminiferous organisms produce a background illumination at least as strong as the daylight percolating through the surface.

Oxygen

The swarms of micro-algae in surface waters photosynthesise carbohydrates during daylight, and at the same time produce more oxygen than they consume in respiration. In hours of darkness they consume more oxygen than they produce. On balance, however, algal activity leads to a surplus of oxygen in surface waters. Below the surface there is no oxygen shortage in the deep oceans; most water masses contain more than enough oxygen for the organisms living there. There is, however, one notable exception: the Black Sea. Like many fresh-water lakes this sea is heavily stratified so that

1·2 Right The planktonic comb jelly *Pleurobrachia pileus* is capable of emitting light. (After J. C. C. Nicol 1964.)

1·3 Below The distribution of light-emitting organs in the families (**a**) Squalidae; (**b**) Paralepidae; (**c**) Sternoptychidae; and (**d**) Myctophidae. The light organs are mainly on the ventral side of the fish. (After D. E. McAllister 1967.)

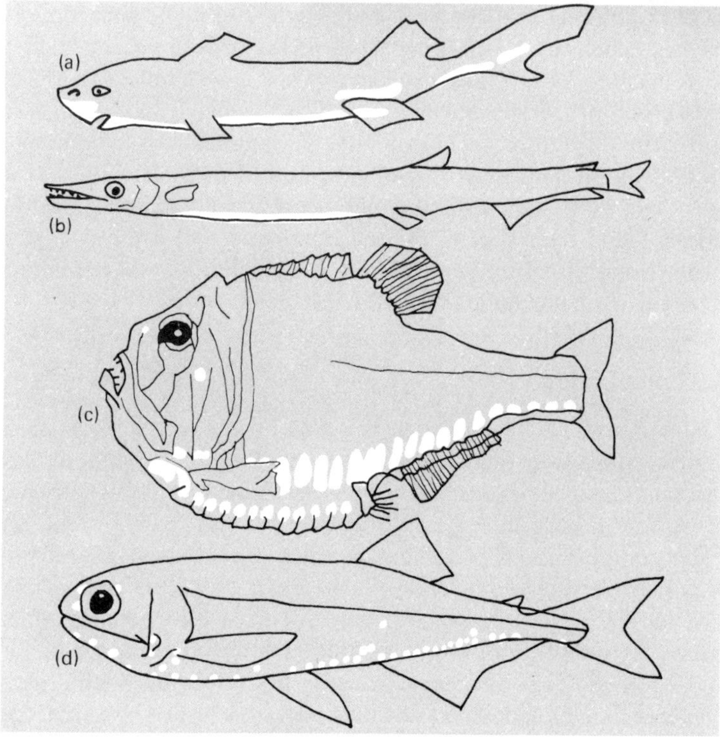

there is no mixing of top and bottom waters. Thus oxygen is found only from the surface to 150–200 metres; at greater depths there is no oxygen and the water is loaded with hydrogen sulphide. Thus the production of plankton as well as of bottom-living animals stops at the bottom of the oxygenated layer and, apart from a rich flora of bacteria, the deeper water of the Black Sea remains barren – a kind of marine desert.

In shallow waters, however, there are situations where oxygen deficiency does occur. If the sea bed is densely populated the bottom 10 to 15 centimetres of water usually has a reduced oxygen content. Oxygen deficiency is also found in the dense beds of eel grass which grow in calm shallow fjords and bays. On a quiet summer morning, shortly before sunrise, practically all the oxygen may have been used up by the respiration of both plants and animals. Mobile animals such as amphipods and fish solve the problem by leaving the eel grass beds at night; but the many sedentary small animals must be able to 'hold their breath' for the time being, or in physiological terms 'build up an oxygen debt' until the eel grass and the micro-algae living on it resume production after sunrise.

Carbon dioxide

As we know, all life on earth depends on the synthesis of plant matter, and the primary ingredients of this 'photosynthesis' are carbon dioxide and water. It follows that there must be a sufficiency of carbon dioxide in sea water, and indeed this is so. In fact, sea water contains about fifty times as much dissolved carbon dioxide as pure fresh water would carry if it were exposed to the atmosphere. The reason is that the sea contains dissolved bicarbonates which create favourable conditions for the retention of carbon dioxide; and while the dissolved gas is roughly in equilibrium with atmospheric carbon dioxide, it is also constantly replenished by the carbon dioxide that is respired by plants and animals. There is thus no shortage of carbon dioxide for photosynthesis.

Hydrostatic pressure

Another interesting property of the sea is the pressure to which organisms are subjected, especially in deep waters. For every ten metres of depth the pressure increases by one atmosphere. It used to be said by biochemists that certain proteins coagulate at a pressure of 850 atmospheres, and it was thought that life could therefore not exist below 8,500 metres. But as we have seen, life exists at 10,900 metres, and at a pressure of over a thousand atmospheres. Moreover, there does not appear to be any special adaptation to depth, because the *Galathea* Expedition not only found a considerable number of bottom-living animals below ten thousand metres, but found also that the species of holothurians, bristle worms and mussels were either identical with or close relatives of species which were plentiful at less than a hundred metres in the fjords of east Greenland.

Actually, it seems reasonable to suppose that marine organisms without any gas filled cavities would be unaffected by pressure, because liquids are only very slightly compressible, and because the cells of such organisms are entirely filled with fluid or semi-fluid protoplasm. Nevertheless this is not the whole story. Dr Zobell's experiments on bacteria from the Challenger Deep seem to show that the pressure under which they normally live is necessary for their full development. Other experiments have shown that a shrimp, *Systellaspis,* which daily undertakes vertical migrations down to about a thousand metres, has a higher rate of metabolism when under increased pressure. But crabs and mussels from the Californian tidal zone which were put under pressure by being lowered in open containers fared very differently: the crabs died at 900 metres and the mussels at 2,250 metres.

1·4 Examples of plankton with appendages which promote flotation. (**a**) The copepod *Oithona plumifera*; (**b**) the coccolithophore *Michaelsarsia*; (**c**) and (**d**) the dinoflagellates *Ceratium*; and (**e**) the crab larva *Sergestes*. (After T. J. Smayda 1966.)

Weight and buoyancy

We have already said that the sea's density enables many creatures to float that would sink in fresh water. But the sea has another important property which even marine biologists often overlook, and that is its *viscosity* or resistance to friction. This varies inversely with the temperature: warm water is less viscous than cold water so that, in tropical surface waters, a body whose specific gravity is greater than that of the sea will sink twice as fast as it would in polar waters. It is essential for phytoplankton to minimise its rate of sinking so as to prolong its life within the photic zone. Also, the swimming animals of the free-water masses profit by increased buoyancy, which enables them to spend less energy on swimming.

Marine organisms tend to be heavier than sea water but many of them have special adaptations to make them neutrally buoyant. Adaptation is to arrange the composition of the body fluid so that it contains ions which are lighter than those of sea water. The protozoan *Noctiluca,* jelly fish, floating fish eggs and some adult fish adjust their specific gravity in this way. Another method is to secrete buoyant substances. Thus organisms such as algae, coelenterates and cephalopods secrete bubbles of gas, while fish are provided with swim-bladders. Another buoyancy device is the storing of fat in the body, either as oil drops (in coelenterates and copepods), or in the liver (as in sharks). Finally, some creatures go to great lengths to increase the body's surface area by means of appendages or the secretion of mucilage. The purpose is to create the maximum frictional surface in order to delay sinking. Some diatoms have very complex and spiky silica skeletons, but they do, nevertheless, sink, if only very slowly. One of them, *Skeletonema*, sinks from 0·30 to 1·35 metres per day; another, *Rhizosolenia*, sinks from 0·16 to 1·77 metres per day. The micro-algae of the tropical surface layers are equipped with much more complicated appendages than are neces-

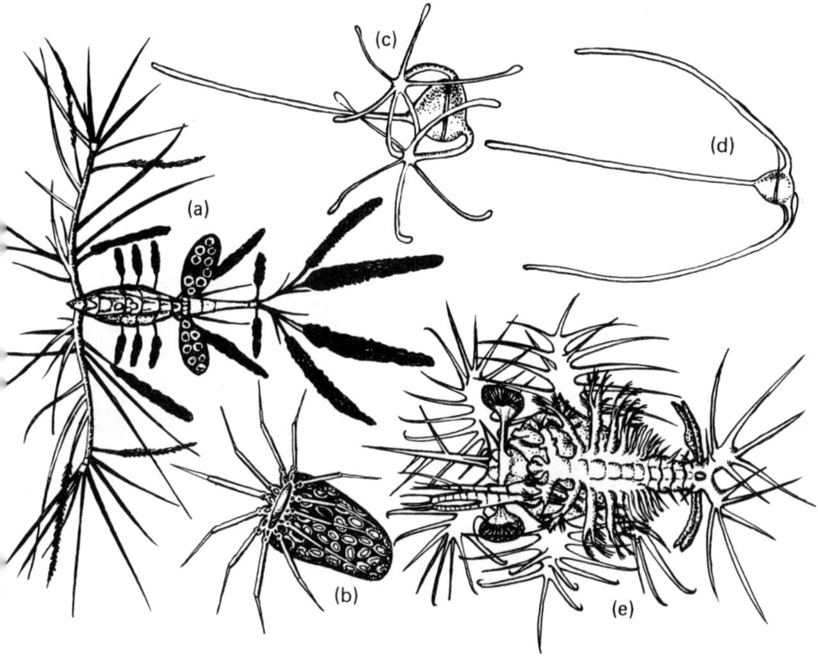

sary for those living in the more viscous polar water. Quite apart from their usefulness in delaying sinking, these complicated structures seem to serve two other purposes. One is the great increase in surface area available for intake of nutrients; the other is the deterrent property of, say, a spiky diatom, which gives it a survival value. It has been observed that some copepods eat spherical and fan shaped diatoms but avoid other species whose shape makes them more difficult to handle.

Suspended matter

It is not only living organisms that float by virtue of the sea's buoyancy; there is also a huge amount of non-living matter suspended in the ocean. There is a collective word – *seston* – which is used to include all bodies living and non-living, floating or swimming in the water. Off Iceland and in the North Sea the non-living component

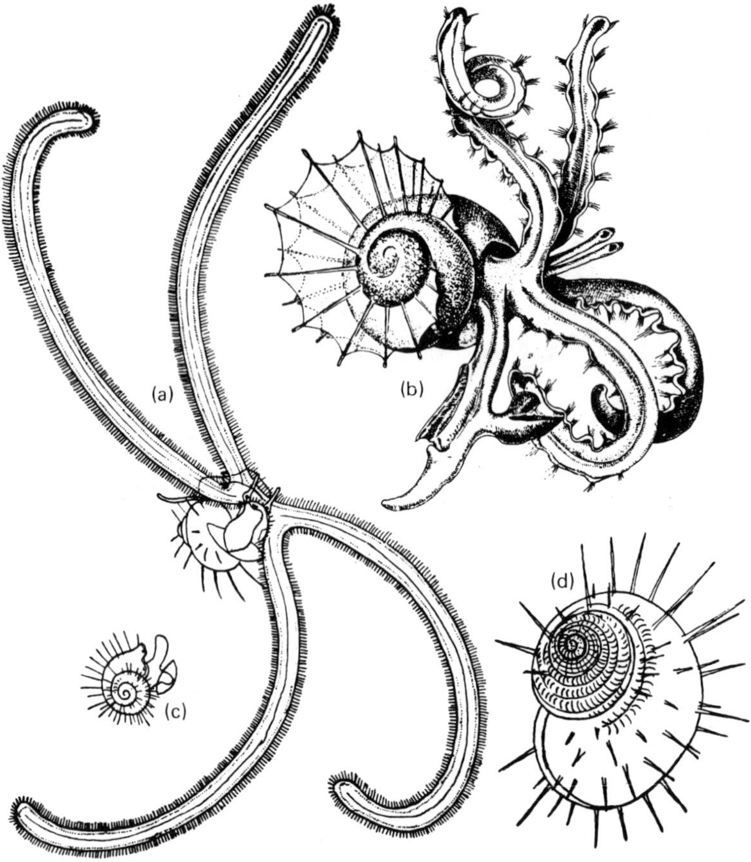

of the seston amounts to between 40 and 90 per cent of the total, even during the peak of phytoplankton production. The non-living component is better known as *detritus* but this may itself consist of two kinds of particles. One kind is inorganic and is made up of clay and mud particles from coastal erosion, silt from melted glaciers, dust from deserts or volcanoes and calcareous or siliceous skeletons of dead micro-algae. The other kind is organic, and comprises dead and decaying particles originating from living organisms. A large concentration of inorganic detritus can indirectly reduce phytoplankton production by increasing the turbidity of the water; but organic

1·5 Larval gastropods that are equipped for long-distance transport in the free-water masses. (**a**), (**c**) and (**d**), larvae of *Tonna* taken off Bermuda; (**b**), larva of an unknown prosobranch from the China Sea. (After G. Thorson 1961.)

detritus, which up to the 1950s was underestimated and neglected by oceanographers, has now proved to be very important in the mechanism of marine food-chains. A rough calculation shows that the amount of organic detritus present at any one time must be at least ten or even a hundred times greater than the annual crop of all marine organisms. Thus, in the north-east Pacific, the total quantity of detritus beneath a square metre of the sea surface appears to be equal to at least 500 grams of dry organic material, and 75 to 80 per cent of this is in the uppermost 100–200 metres. Chemical analysis has shown that this detritus contains a diversity of compounds, which makes it a useful source of food for marine organisms. Furthermore, a large number of marine animals are regularly found with detritus in their digestive tracts. Often, however, this ingested detritus functions as a nutrient for a rich bacterial growth which may prove more beneficial to the animals than the detritus itself.

Algal soup

Having briefly reviewed the physical and chemical conditions of the marine environment, we can see that the primary (vegetative) production must take place in the upper layers. We can also appreciate that the algae which form the bulk of this production must be extremely small if they are to live out their lives in the photic zone before sinking to the bottom. We thus have an immense concentration of microscopic plant life, which in highly productive regions and seasons makes the sea resemble pea soup.

This creates a situation which is almost without parallel on land. It means that the free-swimming larval forms of marine sea animals are literally surrounded by 'instant nourishment'. As soon as they hatch out they have a finely divided 'baby food' ready and waiting to be eaten. Consequently the larval forms need not be highly developed before hatching, and this in turn makes it unnecessary for their parents to lay eggs with more than a minimal quantity of yolk.

We find, then, that many marine animals, which when adult may live at various depths, make a practice of laying their eggs or sending their larvae directly into the zone of algal concentration, in the certainty that the larvae will find abundance of nourishment. In shallow waters, where there is a large bottom-living population, there is sufficient algal production right down to the bottom; and in these waters there are many larval as well as adult forms which simply sit still and have their food brought to them on the tides and currents. Thus the food problem, at the primary consumer level at least, is much less of a problem for small marine animals than it is for land herbivores.

What we have just said might be taken to suggest that the seas and oceans are unfailingly fertile. Actually the fertility, depending as it does on the mixing of currents and on the supply of nitrate and phosphate in surface waters, varies from place to place. It also varies from season to season, again because of fluctuations in currents. It is therefore extremely difficult for even the most experienced biologist to predict next season's fishing potential in a given area. Here we have another contrast with conditions on land. We can map once and for all the barren areas on land, but in the sea such areas are constantly shifting. In general, however, the sea's productivity is more uniformly distributed.

Number and distribution of species

Having briefly reviewed the properties of the marine habitat, we now turn to the degree to which it is inhabited. In theory we should be able to state this in two ways – the size of the marine population, and the number of different species which make up that population. But the size of the population, in other words the number of individuals, is inconceivably great and moreover it fluctuates from season to season. Even if we were able to calculate the population, the number would be so enormous as to be meaningless. We can, however, be more precise about the number of marine species compared with those on land.

There are well over a million species of animals in the world today; but only about 160,000, or 16 per cent, live in the sea. This may be surprising, but we should remember that we are talking about the number of *species* and not the number of *individuals*. In terms of individuals, there is no doubt that many sea areas are just as densely populated as the most populated land areas.

There would appear to be two reasons why there is a seeming paucity of different species in the sea. First, that the relative uniformity of the marine environment has not encouraged the kind of natural selection which on land leads to the evolution of new species. Secondly, that below the surface layers the temperature is low, chemical reactions take place slowly, and again there is therefore less tendency to variation.

Another way of stating the imbalance between land and sea species is to say that the sea has hardly any members of certain groups, such as the insects. On land, at least 75 per cent of all animal species are insects, leaving approximately 250,000 species to be divided between land and sea. On this reckoning the number of non-insect species in the sea represents 65 per cent of the world total.

If we are to describe the distribution of marine species we must subdivide our 160,000 into those that live in the free-water masses and those that live on the sea bed. Those that live in the free-water masses amount to 3,000, or about 2 per cent, of all marine species, and this despite the enormous available living space. The remaining 98 per cent (157,000) live on the sea bed.

In the following chapters we shall deal in more detail with the distribution of animals in different parts of the habitat. It will, however, be helpful to attach labels to the occupants of the different parts of the habitat, as follows:

The name *plankton* comprehends all organisms that drift with the currents. These organisms may be able to make small movements and to migrate vertically, but they are at the mercy of horizontal movements of the water. Plankton subdivides into *phytoplankton* (the drifting plant matter such as diatoms and dinoflagellates), and *zooplankton* (the slightly mobile animals such as small crustaceans, swimming molluscs, jelly fish and free-swimming larvae of bottom animals).

Nekton is the name given to animals that are capable of actively swimming at a speed which enables them to outstrip ocean currents and tidal streams, and even to undertake substantial migrations. In this group we have adult fish, squids and cuttle fish, marine mammals and a few reptiles.

Organisms that live on the sea bed are collectively called *benthos*. There is a subdivision of benthic forms into those that live buried or partly buried in sand, mud or silt, the *infauna*; those that live on submerged or tidal rocks or on the surface of the sea bed, the *epifauna*; and those, such as crabs and prawns, which live at the bottom but can move quickly on the sea bed, the *nektobenthos*.

We shall find that many animals can spend one stage of their lives as benthos and another stage as plankton or nekton, and vice versa. Therefore we must regard these labels as convenient generalisations and beware of attaching too rigid a meaning to any one of them.

2 Plankton

We have defined plankton as consisting of those organisms which are essentially drifters – unable to determine their horizontal position but in some cases able to adjust their location in the vertical plane (just as a balloon skipper may raise and lower his craft to exploit the air currents). By such vertical movements planktonic organisms may obtain or avoid horizontal transportation. They may change their milieu as they develop or they may stick to a certain water mass where life conditions are optimal for them. We have also indicated that many animals can be planktonic at one stage of their lives and yet be part of the *nekton* or of the *benthos* at another stage. This brings us to a further subdivision of the plankton: those organisms that live all their lives in the open oceans (the pelagic zones) are called *holopelagic,* while other organisms that spend part of their lives on the bottom and part as drifters are called *meropelagic*.

Bacteria

We have touched on the importance of plant matter as the source of all food energy through the process of photosynthesis; and we shall have more to say about phytoplankton. But while we are on the subject of marine plant matter, we should deal with *bacteria,* both because these too are often classified as plants, and because they are as essential an element in the marine food-chain as soil bacteria are in the recycling of nutrients on land.

Bacteria have been found in all latitudes, and at every depth examined. In coastal areas we usually find from a few hundred to several thousand living bacteria per millilitre of water; the variation is, however, very wide, from less than ten to more than one million per millilitre. In the open oceans they are less numerous, while in water masses deeper than 1,000 metres we sometimes find no more than a few bacteria per litre of water. Many marine bacteria have the ability to attach themselves to solid surfaces, and in this way their presence is related more closely to the concentration and kind of organic matter than to any other single factor. Since organic matter is produced primarily in the uppermost layers of water, there is, as we have seen, a great deal of detritus and hence a large population of bacteria attaching themselves to and breaking down this detritus. (Much of the surface detritus sinks to the bottom, and here too there is a high concentration, especially on the surface of the bottom sediment.) If we assume an average of 10 bacteria per millilitre in the free-water masses of the open ocean, the total standing crop amounts to about 2,700 million tonnes (1 tonne = 1,000 kg.); this figure would double every 1·25 to 8·3 days, according to species, were it not that bacteria are eaten, or die and sink as fast as they reproduce.

The majority of marine bacteria are apparently beneficial and their role is twofold: they decompose dead organic matter into plant nutrients, and they also convert dissolved organic material into bacterial cell substances which can be assimilated by marine animals from protozoa to those higher up in the food-chain. Bacteria are plainly vital to the functioning of the marine ecosystem, but so far the dynamics of bacterial life in the sea are far from being fully known and understood.

Some species of marine bacteria are, however, detrimental to the economy of the sea or to man's welfare. For example, submerged wooden structures are decomposed by marine microbes. Also, a film of bacteria provides a settling ground for the larvae of animals such as barnacles which foul the hulls of ships; a few species of bacteria may even play a part in the corrosion or deterioration of concrete and iron or steel structures. Finally, bacteria which are not indigenous to the sea, but the result of pollution through the discharge of sewage, can occur in such quantities that bathing has to be forbidden, while infected mussels and other animals may be dangerous to eat.

Phytoplankton

Having briefly discussed bacteria we turn to those microscopic plants for which the name phytoplankton is commonly reserved. It is interesting to note that an Englishmen, Dr J. Vaughan Thompson, invented the plankton net as far back as 1828. It consisted of a bag of silk which he towed through the surface layers catching small larval crustaceans. The plankton net, which now comes in a variety of sizes and meshes, is still one of our most useful devices for catching pelagic organisms; but the mesh of Thompson's net was too coarse to catch micro-algae, and it was not until 1848 that the Danish botanist Anders Sandøe Ørsted discovered their existence while on a voyage to Central America.

Ørsted did more than simply catch phytoplankton; he realised that micro-algae in such enormous numbers were the primary producers on which all ocean life depended. His short treatise on the subject was ignored both in Denmark and abroad for almost fifty years; today it has a place of honour among the twenty most important scientific papers of the nineteenth century. But even Ørsted's net was not able to catch the smallest phytoplankton organisms, and to this day we have not yet succeeded in making a net fine enough to trap the many forms which measure only a few thousandths of a millimetre.

One group in particular, the small naked *flagellates,* has caused plankton botanists many a headache. Too minute to be caught in a net, the scientist must collect a water sample and centrifuge it in order to obtain sufficient numbers. Then when he attempts to count them they dart around under the microscope, making it impossible to tell whether the same flagellate has been counted once or several times. The other groups of marine micro-algae can be killed and kept in alcohol or formalin for subsequent counting, but these naked flagellates literally explode when one tries to 'fix' them. This is extremely frustrating, particularly since we know enough about them to realise that they play an important role in marine ecology.

Recently it has been shown that *heterotrophic* flagellates (species capable of obtaining nourishment from complex organic substances rather than by photosynthesis), may occur in abundance at depths between 1,000 and 5,000 metres in large areas of the North Atlantic. In recent years new counting methods have been tried. One of these is centrifuging followed by a technique similar to that used by doctors when performing blood counts, and this shows prospects of obtaining reliable quantitative results. There are, however, other groups of micro-algae which we can catch with our nets and count in a preserved condition, and these are important enough to examine in detail; they are diatoms, dinoflagellates and coccolithophores.

Diatoms The micro-algae most easily identified are called diatoms. Though extremely small, individual cells are often provided with long appendages, and a row of cells may be joined together in a chain several millimetres long. Diatoms have pairs of siliceous shells which fit into each other rather like a box and lid. Inside the cell wall, the contents divide into two, one part fitting into the 'lid', the other into the 'box'. The two cells formed by the division are thus unequal in size; one is as large as the mother cell, the other smaller. The same process is repeated with each new division, but only up to a certain limit; when this limit is reached both old shells are discarded and the smaller cell doubles or triples its size before forming a new set of shells. With many species this growth to full size is tied to well defined seasons. When the average period between individual divisions is known it can be calculated what size the smallest shells must have at various times.

Armed with this information, planktologists have been able to follow certain water masses by observing the diatom population during its drift across the North Sea and Skagerrak. Diatoms are a very important first link in the food-chain. For example, they are the main food source for immense swarms of copepods which, in turn, are the chief food of small fish such as herring. After death the siliceous shells of the diatoms sink to the bottom where their deposits (diatomaceous ooze) are a characteristic feature of the sea bed, particularly in colder seas.

Dinoflagellates Another very important group of pelagic micro-algae is the dinoflagellates, among which the genera *Peridinium* and *Ceratium* are prominent. Their shells consist of cellulose, and in certain areas and seasons they may outnumber the diatoms, as they can thrive with less nutrient and less light. As food producers they are greatly inferior to diatoms; many animals cannot digest the cellulose shells and, in addition, these shells are often provided with long spines, which make them difficult to swallow. For instance, the common copepod *Calanus* eats large numbers of diatoms, but avoids the dinoflagellate *Ceratium*.

Certain genera, such as *Gonyaulax* and *Gymnodinium,* have a notorious reputation among fishermen. In the tropics in particular they may appear in such numbers that their reddish-brown shells colour the surface and produce the condition known as 'red water' or 'red tide'. The red colour begins to be noticeable when the concentrations of algal cells reaches 200,000 to 500,000 per litre, and the concentrations can rise to 1 to 6 million cells per litre. When the fishermen of these areas see the colouration of the water they know danger is imminent. The algae contain and also secrete poisonous matter, and when they occur in dense concentrations this poison may kill almost all life over large areas. The sad outcome is usually hundreds of thousands of dead fish lying belly upwards on the surface as far as the eye can see. Again, when mussels filter and swallow these algae, poisonous matter is sometimes stored in their livers. (This can happen even in temperate waters and when the

2·1 Some important groups of phytoplankton:
(**a**) Flagellates and blue-green algae;
(**b**) Small Peridinians (*Peridinium*);
(**c**) Diatoms; and (**d**) large Peridinians
(*Ceratium*). (After G. Thorson 1950.)

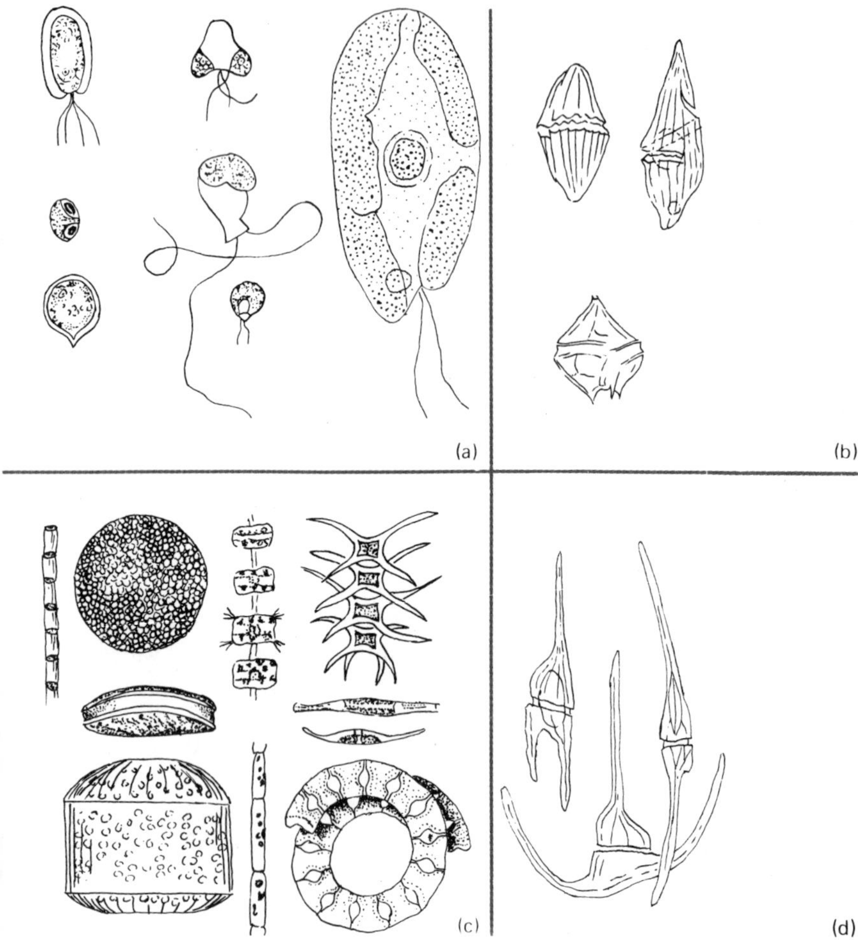

concentration of algae is too small to colour the water.) When sea birds eat these mussels many of them die, and such tainted mussels are also dangerous to man and sometimes fatal.

Coccolithophores Another group of flagellates has a shell composed of a large number of minute calcareous plates, called *coccoliths*, which are a large component of bottom sediments in the warmer seas. It has recently been shown that, after death, coccolithophores are heavy enough to sink to the ocean floor in no more than one to six days – so quickly in fact that they may still contain vitamins when they reach the bottom.

Transitional plankton

While it is proper to distinguish between phytoplankton and zooplankton, it is sometimes difficult to decide where the boundary should be set. The different types of flagellates mentioned above are considered to be plants because they are green and able to produce carbohydrates in the same way as other plants. But there are other groups of flagellates which lack this ability, even though they have the same structure. Moreover many of these are predators, so that we have to regard them as animals. An example is the large flagellate *Noctiluca,* well known for its ability to produce light. It lives chiefly on diatoms but can also make serious inroads into swarms of zooplankton such as mussel larvae. There are also two groups of single-celled organisms which must be included in the animal world. These form a transition to the 'real' zooplankton.

Foraminifera are an important component of free-water masses. Curiously enough, and in spite of their abundance, only about thirty species of pelagic foraminifera are known so far; of these twenty-two species are restricted to the warm oceans between 40°N and 40°S latitude, while the few cold-water species are bipolar. Foraminifera seem to spend their earlier stages in the photic zone, descending later to deeper waters. They are mainly predatory, catching their prey by means of a network of extruded protoplasm. Their shells are calcareous: some have cavities and others are lattice-shaped, often with very beautiful patterns. The most important of the pelagic forms is probably the genus *Globigerina,* whose shells form thick layers of bottom deposits (globigerina ooze) on the ocean floor. Lastly the radiolarians, a group that feeds mainly on unicellular micro-algae and protozoans, have shells that are veritable works of art, with beautifully ornamented lattice structure consisting of silica and strontium sulphate. Here again, the shells of the dead animals sink to the bottom.

Ocean sediments Huge areas of the ocean bed are covered with thick layers of sediments built up of the skeletons of planktonic micro-organisms. It is the calcareous skeletons which usually predominate in the tropical and warm temperate regions. These can be subdivided into globigerina ooze, coccolith ooze, and pteropod ooze (from small swimming molluscs), collectively covering an area of about 128 million square kilometres. The skeletons of siliceous micro-organisms forming diatomaceous and radiolarian ooze predominate on the bottom of temperate and polar seas, and cover an area of 38 million square kilometres. A further 102 million square kilometres is covered with red clay of inorganic origin.

Zooplankton

Having briefly dealt with those forms which are difficult to classify as plants or animals we now come to that part of the total plankton which is definitely animal – the zooplankton. As we shall see, the word zooplankton comprehends forms ranging in size from 0·01 millimetres to several metres; but its prime importance in the

2·2 The comb jellies *Beroë* (top, centre) ; and *Pleurobrachia* (bottom middle) ; the bell-shaped medusa *Aglantha* (middle right) ; the slender arrow worm *Sagitta elegans* (top right and bottom left) ; the polychaete *Tomopteris* (centre) ; and (below this to the right) the larval cephalopod *Eledone.* Scattered among them, various fish and crustacean larvae. (After J. Fraser 1962.)

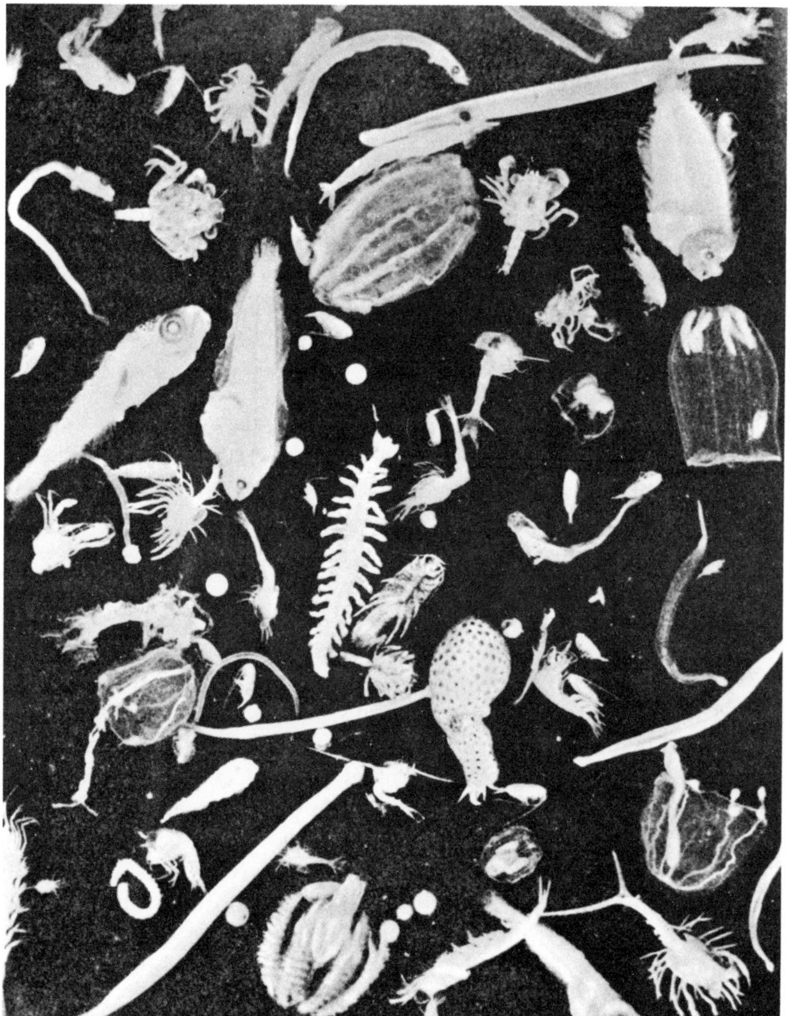

economy of the sea lies in the fact that most zooplanktonic organisms convert the vegetable matter of the phytoplankton into animal protein, that is, they are key-industry animals. The smallest planktonic animals stand in the same relationship to algae as, on land, cows do to grass. They are the main herbivores of the sea.

But the zooplankton also includes many other and larger forms which prey on the herbivorous members, so that predator and prey coexist. Herbivorous animals may be eaten either by larger plankton animals or by the active swimming animals of the *nekton* (chapter 3). Either way, the zooplankton complex is the bridge between the microscopic pastures of the sea and the largest marine creatures.

Outstanding among the converters of vegetable matter to animal protein are two groups which also may well be called 'key-industry' animals; these are the copepods and the euphausiids. These animals are so efficient at sieving phytoplankton (a task which man is as yet unable to perform economically), and exist in such colossal numbers, that their importance as a possible food source for man as well as for marine animals must be stressed.

Copepods There is little doubt that the swarms of copepods in the sea constitute the largest single group of protein producers in the entire world. In the north-western Pacific there are an average of 15,000 copepods per cubic metre of surface water, while off the coast of Murmansk a single species may number up to 28,000 individuals per cubic metre. Even at depths of 500 metres their abundance is remarkable – up to 1,500 individuals per cubic metre, as off the coast of Franz Joseph Land.

The average weight of zooplankton in the North Sea is 10 million tonnes (wet weight), of which copepods constitute at least two-thirds. Such quantities are still more impressive when we realise that most zooplankton organisms have several generations per year; a large North Sea copepod such as *Calanus finmarchicus* has two to three generations, and smaller copepods have several more. A single pair of one very small species, *Tisbe furcata*, with a volume of only 0·14 cubic millimetres, can produce four generations in one hundred days; if all the offspring survived and propagated within the same period, the first pair would be progenitors of 1,055,000,000 individuals with a total volume of 70 litres.

The smallest copepods are only about 1 millimetre in size; the largest are as big as a grain of rice (about 5 millimetres long). *Calanus finmarchicus* is one of the largest and is perhaps the most important single food animal. It is almost cosmopolitan in its geographical distribution, and is found from the sea surface to 4,000 metres. In fact the British zoologists S. M. Marshall and A. P. Orr have written a whole book on this one species. From them, we know, for example, that the amount of food available and the water temperature completely govern its rate of propagation. Off East Greenland it takes a whole year from the time the egg is hatched for the species to reach the spawning stage; off west Norway the entire growth takes three months, and in the English Channel it takes only two months. We know that individuals in temperate seas live at most seven to eight months, but that they are usually eaten at a much younger stage. We know that in volume, 658 adult females and 830 adult males make up one cubic centimetre, and that they are at their fattest in the spring and most meagre in the winter. We know fairly accurately their dry weight, vitamin content and chemical composition.

We know, too, how many diatoms they eat, and that the male eats much less than the female but is better able to utilise its food. We know that the necessary daily ration of food for the older larval stages is between 50 and 70 per cent of the body weight, that the *Calanus*-type have a complete turnover of their intestinal contents from four to six times a day and that they assimilate their diatom food much better when they graze in low rather than in high

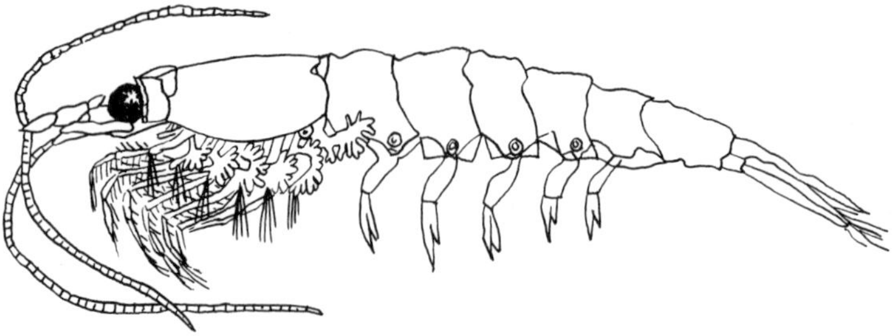

concentrations of micro-algae. We know also the most important enemies of the species, and that they can be attacked by at least twelve different parasites.

Our knowledge is, however, still very limited considering the important position which *Calanus* occupies in the marine food-chain, as well as the steeply increasing part it will play in future calculations on food production and hence, indirectly, on the food resources of man.

Euphausiids The other key-industry member of the zooplankton may prove in the long run to be even more important to man than the copepods. Whereas even the largest copepods (at 5 millimetres) are too small to be caught profitably, the euphausiids are much larger, ranging in length from 2 to 5 centimetres. They are shrimp-like in appearance and have a wide distribution, especially in Antarctic waters, from the surface down to 5,000–6,000 metres. Euphausiids carry light organs; each species can, at will, display its own particular light pattern. If, for example, specimens of *Maganyctiphanes* are placed in a row of glass aquaria in a darkened room, and if the specimens in the first aquarium are stimulated to produce light, there is an immediate response from the members in the next aquarium, and so on, in a chain reaction.

Euphausiids are basically herbivorous, like copepods, but they are also carnivorous when the opportunity offers, preferring copepods to algae. They also have a much longer life than copepods. In the Antarctic and in cold temperate waters it takes two years from the time the larvae hatch until they are sexually mature, and they die shortly after spawning. In the tropics the life-cycle takes one

2·3 The euphausiid *Meganyctiphanes norvegica,* common over all the North Atlantic. (After G. Thorson 1961.)

year, but quantitatively the tropical species play a much smaller role and do not appear to swarm on anything like the scale of the cold-water species.

In the Antarctic the largest species, *Euphausia superba,* occurs in inconceivably large swarms. Whalers call them 'krill' and this euphausiid is almost exclusively the food of the enormous baleen whales, which until recently were particularly numerous in these areas. A few figures will illustrate the amounts of food being discussed. In 1955–6 alone, about 40,000 baleen whales were caught, of which at least three-quarters were in the Antarctic. These large whales need so much food that the energy expended in actual hunting would exceed the returns. However, they are adapted to obtaining nourishment simply by straining krill from the sea as they cruise slowly along, and a large baleen whale will eat an average of about 850 litres of euphausiids per day. Of course, there are many more baleen whales in the Antarctic regions than those killed in a single year, and there must be very many more euphausiids than those accidentally finding themselves in a whale's mouth. Accordingly a recent calculation suggests that *Euphausia superba* alone might yield an annual output of about 100 million tonnes – twice the yield of today's world fisheries. Furthermore, euphausiids contain more vitamins than most other crustaceans. Their fertility is not too impressive (less than 10,000 eggs per female) but their survival rate seems to be high.

Like all other crustaceans, euphausiids grow by moulting. They cast off their skins after a new and more spacious one has been formed beneath. Laboratory experiments by Reuben Lasker on *Euphausia pacifica* have shown that it moults every five days. This means that the weight of skins moulted equals its own body weight every fifty days, or seven times its own body weight per year, the moults sinking an average 300 metres per day. If the animals moult in nature at the same rate as in the laboratory, euphausiid moults must

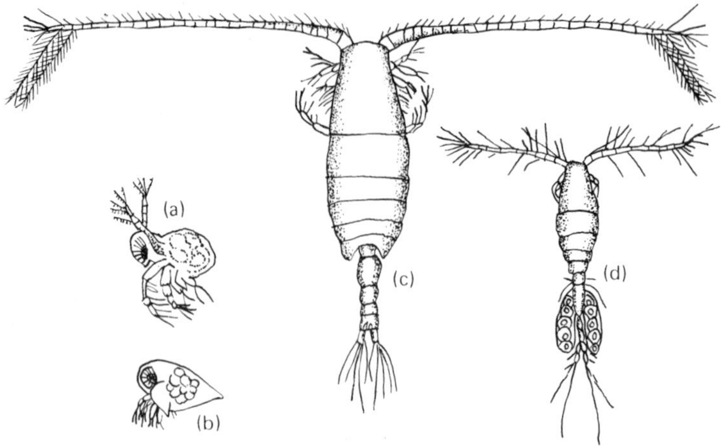

constitute an impressive fraction of the total rain of detritus on to the sea bottom.

We have paid a good deal of attention to copepods and euphausiids, and rightly, because of their importance in marine food-chains. Nevertheless, we must remember that in the zooplankton of the world's seas and oceans there are representatives of almost every phylum of the invertebrates. Vertebrates do occur in the plankton, but only temporarily, as in the case of larval plaice, drifting in the surface waters in their early stages before settling down to an adult life on the sea bed. We have already defined such creatures as being *meropelagic,* and usually their planktonic phase lasts only a few weeks. There is, however, one vertebrate which spends three to four years drifting with the Gulf Stream for more than two thousand miles, from the Sargasso Sea to the coasts of Europe: this is *Leptocephalus,* the transparent larval form of the common eel (*Anguilla*).

Zooplankton consists for the most part of quite small animals, but they range in size from the 1 millimetre copepod mentioned above to the common sea nettle (*Cyanea*), a jelly fish with a diameter of anything up to two metres, and with eight-hundred stinging tentacles trailing behind as it pulses its way slowly through the water.

Pleuston Two further subdivisions of zooplankton are to be noted – the pleuston and the neuston. The first refers to those animals that

2·4 The cladoceran water fleas (**a**) *Podon* and (**b**) *Evadne,* originally fresh-water crustaceans which have adapted to salt water; and the copepods (**c**) *Calanus finmarchicus* and (**d**) *Oithona similis.* (After G. Thorson 1961.)

not only live at the surface but protrude part of their bodies into the atmosphere. Prominent members of this subdivision are the *siphonophores,* colonial coelenterates whose bodies are over 95 per cent water, and almost impossible to handle, since their gelatinous substance is very fragile. Some of these form colonies in which the uppermost individuals stand up above the water surface, the most interesting of these being the Portuguese man-of-war, *Physalia caravella,* a colony composed of single individuals with widely different functions. Above the water the colony is represented by a blue-violet, balloon-like, air-filled floating bladder, about the size of a clenched fist. When floating this bladder looks very like a wind-filled sail on an old warship. And when, as is often the case, one meets a whole mass of these floating bladders moving around on the surface, the impression is of old caravels in full sail. Sometimes these 'fleets' are so numerous and close together that together they can cover an area of at least one square kilometre. From the bladder the rest of the colony's individuals hang down beneath the surface; these are subdivided as to function into the rather inconspicuous reproductive individuals, the funnel-like white nutritive individuals (which are actually only large mouths) and the defence individuals (threads several metres long with stinging cells). These stinging threads are feared by humans along the shores of the Mediterranean and almost all tropical waters. If a child happens to be stung by one of them, the result may be fatal. Children up to twelve years of age have died of the shock. If an adult swimmer is stung he will usually survive but the experience can be quite frightful. The poison of the nettle cells works instantly on contact, and it is just as though one had received an electric shock. Other symptoms appear quickly: an almost unbearable pain in the area of the sting, swelling, intestinal pains, cold sweat, headache, vomiting, and sometimes even paralysis and collapse. As a rule the victim feels extremely weak and must be confined to bed for some days.

But *Physalia* has its own enemies. One of these is the pretty blue-shelled pelagic snail, *Janthina,* about 2–3 centimetres long, which trails along the surface bound to a gas-filled float of mucous bladders. Two medium-sized janthinas can devour a *Physalia* colony with a 10 centimetre long floating bladder in twenty-four hours. Sea turtles can also take *Physalia* without being harmed by their tentacles.

Another siphonophore which is far less dangerous has the sailors' name of 'Jack sail-by-the-wind' (*Velella velella*). The float consists of a flattened oval about 5 centimetres long, on which is mounted a triangular fin enabling the colony to sail across the wind in the same way as *Physalia.* Both these siphonophores are *dimorphic,* that is, those of the Northern and Southern Hemisphere are mirror images of each other; so that their sails are trimmed to the prevailing winds, which also are contrary to each other.

The Queensland sea wasp medusa, *Chironex fleckeri,* is probably the most dangerous animal of the sea. Although only a few inches in size, its sting can kill an adult man in a few minutes. During the last twenty-five years sharks claimed only thirteen victims in this area while sixty persons were killed by sea wasps.

One more group of coelenterates, the comb jellies (ctenophores) are known for their voracity. A West Atlantic genus, *Mnemiopsis,* is often found at a density of fifteen individuals per cubic metre, and they can catch and digest some 22,000 oyster larvae per day.

Salps, which are transparent, gelatinous tunicates, may occur in up to 2,500 individuals per cubic metre in the tropics. They can put large steamers out of action by completely clogging their water-cooling systems.

Neuston A community which inhabits the topmost few centimetres of the water has until recently been overlooked by the biologists, probably because special gear is needed to catch its members. This

2·5 The Portuguese man-of-war *Physalia caravella*, which has captured a fish, paralysed it with its stinging cells, and has partially digested it. (After G. P. Wilson 1947.)

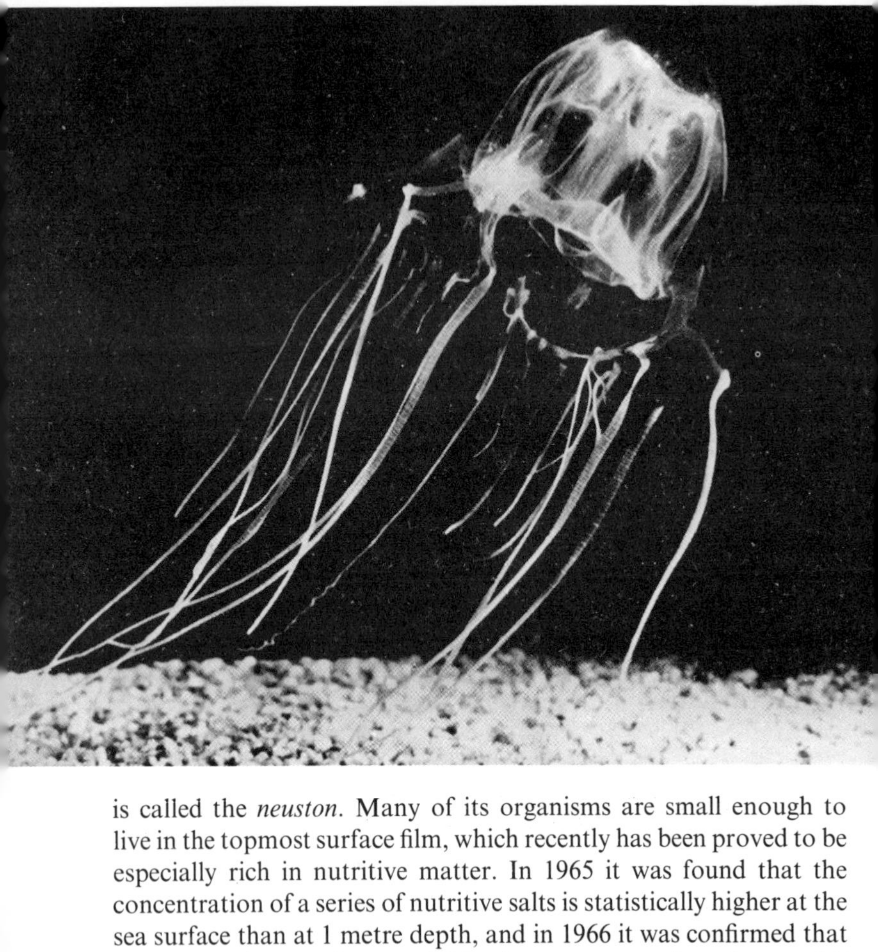

is called the *neuston*. Many of its organisms are small enough to live in the topmost surface film, which recently has been proved to be especially rich in nutritive matter. In 1965 it was found that the concentration of a series of nutritive salts is statistically higher at the sea surface than at 1 metre depth, and in 1966 it was confirmed that there is more organic material in the uppermost sixty-thousandth of a millimetre than at 10 centimetres. Consequently, this narrow stratum contains from 10 to 1,000 times more bacteria than the water immediately below. Seven new species of bacteria were found, and these must be resistant to ultra-violet light, which is lethal to most bacteria. They are themselves eaten by large numbers of protozoans. Other creatures in this crowded layer are the larvae of bottom invertebrates and copepods, floating fish eggs and larval

2·6 The sea wasp medusa *Chironex fleckeri* is probably the most venomous animal in the sea. (After W. M. Stephens 1966.)

fish. Many oceanic birds have bills adapted for skimming this layer, and we know of tropical bats which take their food from the neuston.

In the tropical neuston we also meet one of the astonishingly few groups of insects associated with the open sea: these are the water-striders, *Halobates,* which are closely related to the fresh-water forms occurring on lakes and rivers. They are so light that they can use the surface as a skating rink, and they can move at an impressive speed. They are not strictly marine animals, since they live on the surface (the epineuston) of the sea and deposit their eggs on flotsam such as drifting feathers. They cannot survive when submerged in sea water.

The population of the neuston also includes animals that are associated with driftwood and floating sea weeds. These are species which need a solid base on which to settle or crawl, and which, therefore, can be regarded neither as pelagic nor as bottom-dwellers. Typical of this way of living are the goose barnacles, *Lepas*. This is a genus of primitive crustaceans, whose free-swimming larvae attach themselves to flotsam. They obtain food by combing the water with limbs transformed into a brush-like organ. This floating existence makes them wholly dependent on surface currents as a means of transport, and they are therefore widely dispersed. Several of the species are cosmopolitan and are found in all great oceans. This dispersal has been greatly aided in the past by the long slow voyages of sailing ships.

Sargasso flora and fauna

One completely unique fauna, found only at the surface of the Sargasso Sea, depends for a habitat on the drifting seaweed of the Sargasso Sea. All true plankton algae are microscopic and cannot form attachment areas for larger animal forms, but the Sargasso weed is

2·7 Below The goose barnacle *Lepas fascicularis* adhering to a bladder of the brown algae *Ascophyllum* (left), and suspended by its own float (right). (After E. Thörner and W. E. Ankel 1966.)

2·8 Right Floating plants of *Sargassum* inhabited by the crabs *Portunus sayi* (top right), and *Planes minutus* (lower left) ; the fishes *Antennarius marmoratus* (centre), and *Syngnathus pelagicus* (lower right) ; the shrimps *Palaemon tenuicornis* (middle left) and *Latreutes fucorum* (bottom) ; the snail *Scyllaea pelagica* (a little below centre) ; the polychaete *Platynereis dumerilii* (bottom) ; the goose barnacle *Lepas pectinata* (bottom right) ; hydroids (top left, enlarged) ; a small snail and a sea anemone. (Kai Olsen.)

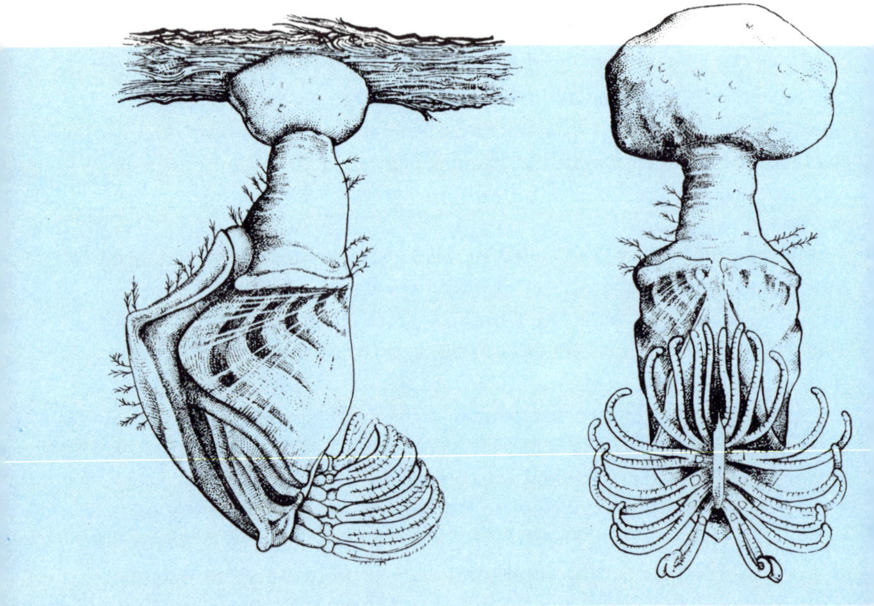

composed of two species of attached brown algae, *Sargassum natans* and *Sargassum fluitans,* which normally live in the shallow waters of the tropical Atlantic coasts of America. When these weeds are detached by the force of the waves, they float because of their air-filled bladders, and follow the current to the Sargasso Sea. Here the weed is trapped at the surface of an immense whirlpool

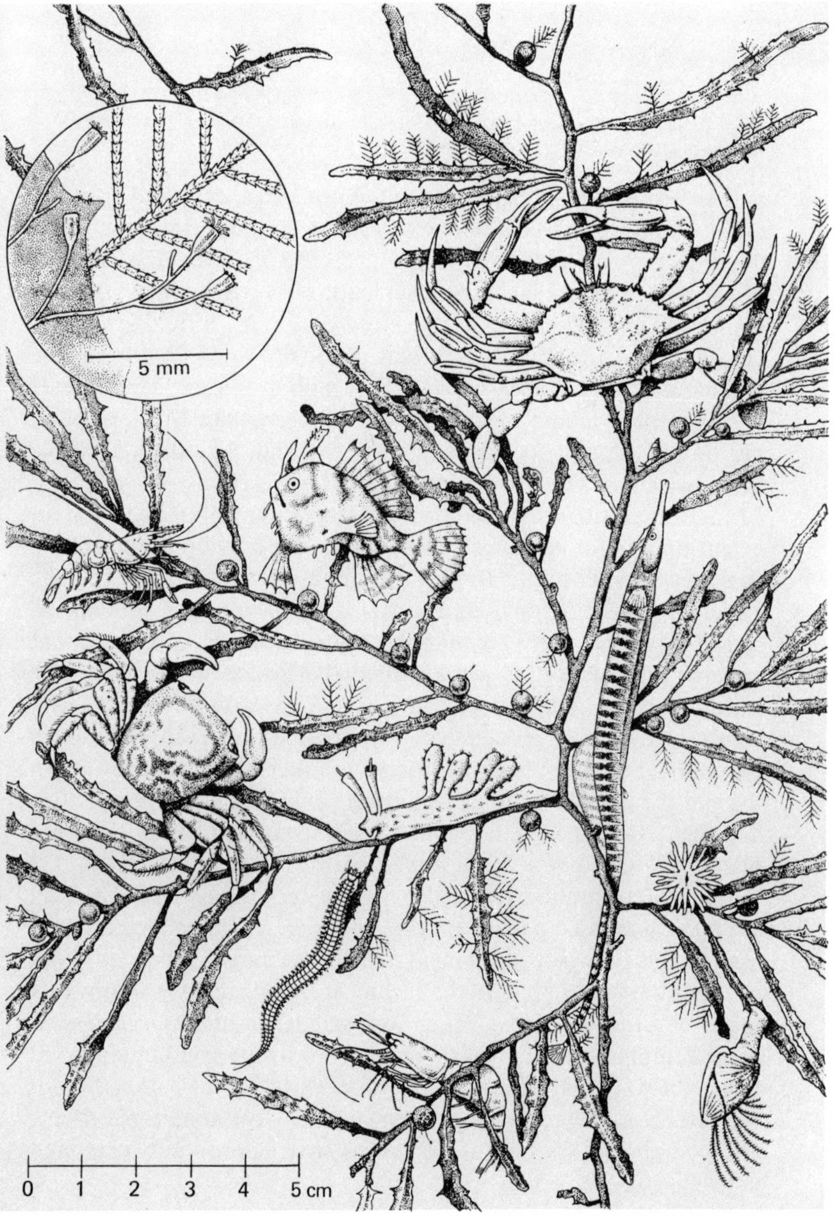
5 mm
0
1
2
3
4
5 cm

and is 'stored' in the surface itself. Even when detached from its usual habitat the weed can live on and continue to grow while floating, though unable to reproduce. The most detailed studies of Sargasso weed to date have been made by two Scandinavians: the Dane, Øjvind Winge, and the Norwegian, A. Eide Parr. They calculate that the total wet weight of all sea weed floating in the Sargasso Sea is between four and ten million tonnes. This sounds impressive but in fact it does not average more than 1 to 1·5 tonnes per square kilometre. It is quite scattered, and old stories of whole fleets of sailing ships being immobilised are certainly exaggerated. From time to time the weed drifts in large clumps but hardly to an extent that might be dangerous to ships. Not only does the moving Sargasso weed continue to grow, but individual plants can become extremely old. For obvious reasons it has never been established just how old, but there are marine scientists who are convinced that it is still possible to find algae in the Sargasso Sea which lay on the surface when Columbus passed through the area in 1492. There seems, however, to be a tendency among these algae to a decline in the number of vesicles or bladders with increasing age, so that in the end the weed is no longer buoyant, sinks below the photic zone and dies. Nevertheless individuals seem to survive so long that old plants far outnumber the new plants torn off the coasts. Thus, the *Sargassum* accumulations really form a community of their own.

Practical minded Americans did, of course, examine the potentialities of this immense algae mass a long time ago, but the collecting of the sea weed as well as the technical treatment of it afterwards raises serious problems. The 'wet-weight' of the *Sargassum* is indeed impressive, but 1,000 tonnes of brown algae contains 830 tonnes of water as well as material for which no use can be found. Nevertheless, the technicians seem not to have abandoned hope, and perhaps the reserves of *Sargassum* may be profitably exploited in the future.

A unique animal life is found on and among this floating weed. More than fifty species have been counted, including hydrozoans, actinians, polychaete worms, snails, shrimps and crabs. Many of these are dwarf versions of animals originally transported from coastal areas. The permanent sedentary *Sargassum* of the coastal zone is much richer in associated animals than that of the floating weed species, and the composition of these two communities is quite different.

There are however special 'Sargasso weed forms', the most interesting being the little toad fish, *Antennarius marmoratus*. With its white, gold and brown-marbled exterior, and with fins and other appendages protracted into jagged lobes, it is hardly distinguishable from the sea weed in which it lives. It is a voracious predator, its favourite food being shrimps, which are usually swallowed tail first. In this way the predator avoids the trouble of the long, sharp rostral spines. *Antennarius* may even attack and swallow fish double its own length. Other fishes characteristic of the *Sargassum* weed fauna, are a small sea horse, *Hippocampus*, and a pipe fish, *Syngnathus*, which also are well camouflaged both by their shape and their colour, as is the small Sargasso crab, *Planes minutus* – a jigsaw puzzle of camouflage colours.

3 Nekton

We now come to that component of marine life which is capable of swimming purposefully and powerfully, at speeds which make them independent of ocean currents. The nekton consists almost entirely of vertebrates: indeed the only invertebrates that are fast swimmers are the torpedo-shaped squids.

Squids

Squids of the genus *Loligo,* whose members range from 10 to 20 centimetres in length, often occur in huge numbers, and are sometimes trapped in seine nets and mid-water trawls. In northern waters they are not very numerous, but in 1958 there was a massive invasion of squids into the fjords of northern Norway; the resultant catch weighed 8,500 tonnes and fetched nearly half a million dollars.

In general the largest squids have eluded capture. It is worth mentioning the largest squid of all, *Architeuthis,* whose body length plus the tentacles can measure up to sixteen or seventeen metres. They are sometimes seen at the surface, where they may have given rise to some of the sea serpent legends. Others have been discovered only because an occasional dead body has been washed up on the coast. They move far too quickly to be caught in nets, but it has been known for many years that they are very numerous, because they are the main food of the sperm whales (*Physeter catodon*). These whales are themselves still very numerous, so much so in fact that an annual exploitation of eighteen thousand animals does not appear to diminish their numbers. They are voracious and like all true predators have a rapid digestion. It follows, then, that large squids must be much more numerous than was originally supposed, and this is corroborated by the fact that with the advent of 'big game' fishing, it is often possible to catch squids with long baited lines, or (off the Pacific coasts of South America) with a rod and line as used for tunny fishing. Recently a British experiment, using a baited line together with a camera and flashbulb, revealed that another large species (*Ommastrephes pteropus*) is extremely common in the North Atlantic and can be caught as often at the surface as at 100 metres. Some of these large squids can swim so fast that they are able to shoot right out of the water, and have even been known to land on a ship's deck.

Fish

By far the most important component of the nekton are fishes, and even though the number of species of pelagic fish is not very large, they play an important role in the turnover of the sea, especially when seen in relation to food for man. About 50 per cent of all fish caught for human consumption in 1966–7 belonged to the herring family, and a further 10 per cent to the mackerel and tunny families.

Many pelagic fish, especially those living near the surface, are very fast swimmers. Just how quickly ~~they swim~~ has been the subject of experiments in direct observation by marking a large number of individuals and later recapturing them. Laboratory experiments are difficult, partly because the size of experimental tanks is necessarily limited, and partly because one can never be sure that the fish will swim at the same speed as under natural conditions. Also, it is impossible to know how the experimental speeds compare with the cruising or endurance speeds. Actually, the speed of a fish when avoiding an enemy or a net – the so-called *burst* speed – can be maintained for only a very few seconds.

Let us take a look at some figures which most modern scientists consider to be reasonably accurate. There would appear to be a direct relationship between the size of the fish and the speed it is capable of. For reasonably fast swimmers the rule seems to be that a fish can move approximately ten times its own body length in one second. For the large fishes this means impressive speeds and it seems as if some of the very largest species can swim even faster – the tunny and its relatives can reach up to 13·4 times their body length per second. Underwater observations on 'skipjack tunas' in giant tanks have indicated cruising speeds of 230 to 360 kilometres per day. Two large tunny, tagged in the Bahamas, were recaptured near Bergen in Norway less than four months later, which corresponds to a cruising speed of between 65 and 162 kilometres a day. There is no doubt that the tunny is particularly suited to fast swimming: of all fishes it probably has the most elegant and streamlined body, and even the dorsal, pectoral and anal fins can be folded into a groove when the fish is swimming at speed, thus offering no resistance whatever to the water. However, while being marvellously designed for high speeds the tunny is also an example of how much it costs to force the speed over a certain limit – something ship builders are well aware of. Alone among sea fish the tunny has red flesh, and this alone is a sign of high muscular activity. Such activity is confirmed by the fact that the body temperature is between 3°C and 12°C above the temperature of the surrounding water. Unlike the true warm-blooded sea animals (whales, seals and sea birds), the tunny is not insulated by a solid layer of fat under the skin, so it must suffer a considerable heat loss; and the effort of maintaining the high temperature must cost a lot of energy and call for a very high metabolic rate. Accordingly, young tunny have to eat 25 per cent of their own weight every day in order to grow normally. The tunny seems to have evolved into a 'vicious circle' situation in which it has to have a high metabolic rate in order to range over wide areas in pursuit of concentrated shoals of other fishes with which to sustain its high metabolic rate.

As far as speed is concerned the tunny is in a class by itself. Another fast swimmer is the salmon, which can be marked in rivers when young, to yield reliable information when later recaptured. The average cruising speed for migrating salmon is recorded to be from 27 to 36 kilometres a day (Scotland) to 38 to 64 kilometres a day (Norway). One salmon, 85 centimetres long, travelled along the Norwegian coast, a distance of 1,100 kilometres, in about eleven days or less, an average cruising speed of at least 100 kilometres a day, since it would not necessarily travel in a straight line. Most other fish, even surface fish, are considerably slower. Nevertheless, the herring has been found to maintain a cruising speed of 78 kilometres, the cod of 65 kilometres, and the European eel of 24 to 72 kilometres a day.

Among the fishes of the nekton, we know many other feeding types besides the rapid super-predators. The herring, for instance, usually feeds on copepods, which are snapped up individually. The large, clumsy sun fish, *Mola*, which can weigh up to a tonne, has a tiny mouth and feeds mainly on jelly fish, pteropods and on the small crustaceans in the surface layers. None of these fish need to swim particularly fast.

Schooling The survival value of schooling among fish has recently been investigated and the results are worth mentioning. A school may be defined as an aggregation of fish in which the spacing and orientation of the individuals is uniform. We know of thousands of species of marine fish that form schools, schools which may vary from a few individuals up to enormous shoals. (In the North Sea, herring schools 15 to 17 kilometres long and 5 kilometres wide have been seen.) Schools usually consist of fish that are always preyed upon or that are preyed upon during the major part of their life. Many of the latter belong to the herring and mackerel families – fish that are pelagic and swim fully unprotected in the open sea. Not all fish schools have the same structure, but some behavioural patterns are typical.

Experimentally, grouped fish have been found to have a strong

tendency to school. They are reckless when separated, but when schooling they exhibit less movement, consume less oxygen, learn faster, retain impressions longer, and eat more than do isolated fishes of the same species. Even fry of schooling fish raised individually in isolated jars will school as soon as they are brought together. A school fish is strongly attracted by its own mirror image and will hardly move out of the field of vision of its reflection. Schooling fish separated by a glass wall move towards each other and school on either side of the glass. Each school usually consists of one species only, though exceptions are known, and all the members of a school are about the same size, probably because large fish swim faster

3·1 Fully formed ball of young schooling fish (*Sebastodes paucispina*) found under the stern of a small boat. (After C. M. Breder jr. 1959.)

than smaller ones. Very large schools of herring may, however, comprise fish of different sizes. Permanent schools are extremely stable and are not broken up unless there is a violent disturbance. When there is an alarm, which is usually followed by the voiding of alarm substances from their bodies, they pack together and normally point towards the source of disturbance. Even an attack by sea birds makes them group more readily, but while schools are less concentrated when its members are hunting for food, schooling fish seem to swim almost noiselessly or at least producing so little sound that a predator discovers the school by seeing it long before it is able to hear it.

A school in which all individuals respond in the same way and as a unit seems in most cases to have no permanent leader. The ideal shape of a school trying to make itself inconspicuous would be spherical, with the largest number of fish presenting the smallest surface area, but migrating fish often form ellipsoid schools or some other pattern, even in some cases lining up in parallel rows.

The American fishery biologist E. S. Iversen has summarised this important topic so well that I can do no better than to quote him:

> Findings suggest that if a fish is a member of a school, it has a better chance of living longer than if it swims alone, because predatory fish searching for food have less chance of encountering a school than of encountering individual prey fish scattered through wide areas of the ocean. A predator will encounter a school of prey fishes only occasionally. Since it can eat to capacity only, the larger the school the less the probability of any individual member being eaten. Let us assume ten fishes is the most a certain predator can eat. In a school of one hundred prey fishes, the chance each has of being eaten is one in ten, but if the school consists of one thousand fishes, then the individual chance of being eaten is reduced to one in a hundred. If, on the other hand, a predator encounters individual prey fish rather regularly, though at widely spaced intervals, it would never become satiated, and every fish encountered would be consumed.

3·2 Top Flying fish beginning their 'flight', with the V-shaped trail left by the tail on the water surface. **Bottom** Flying fish when airborne. (After G. S. Carter 1945.)

It may be added that the chance of actually avoiding discovery will be greatest for small schools, though the addition of a few extra fish to a small group will make the conspicuousness of that group increase at a much lower rate than the number of its members. Here the main advantage seems to be that the schools are difficult for predators to detect. But in very large schools the advantage may be quite the reverse. Such huge groupings packed densely together may well act on a predator as a 'monster' and frighten it away.

Observations and experiments have clearly shown that the main sense keeping the individuals of a school together is vision. They always line up so densely that they can easily see each other's responses, and a school with eyes all over its periphery may well detect an approaching predator more efficiently than a single fish. We also know that blind fish and fish in total darkness do not school. Since sight is the most important factor in the schooling process it follows that schools will disperse or at least loosen at dusk and re-school at dawn. In some cases we know, however, that schools do not completely disintegrate at night, probably because their ability to smell other individuals and to pick up sound waves by means of the lateral line sense organs helps them to keep together without being in a real school (i.e. a non-oriented group). This breakdown of the schools at dusk is a most important factor. Schooling fish are often feeding voraciously just at dusk, which means even more dispersal, so that predators may find and attack one individual fish at a time. Let us suppose that a school of herring dispersing at dusk is simultaneously attacked by a group of cod and a group of pollack. We know that a cod perceives its prey at a light intensity ten times smaller than a pollack; this will give the cod a much better chance than the pollack to hunt herring during the critical period of fading light. Furthermore, if a predator can see its prey for just fifteen minutes longer than the prey fish can see the predator, this must strongly influence the balance between the two species.

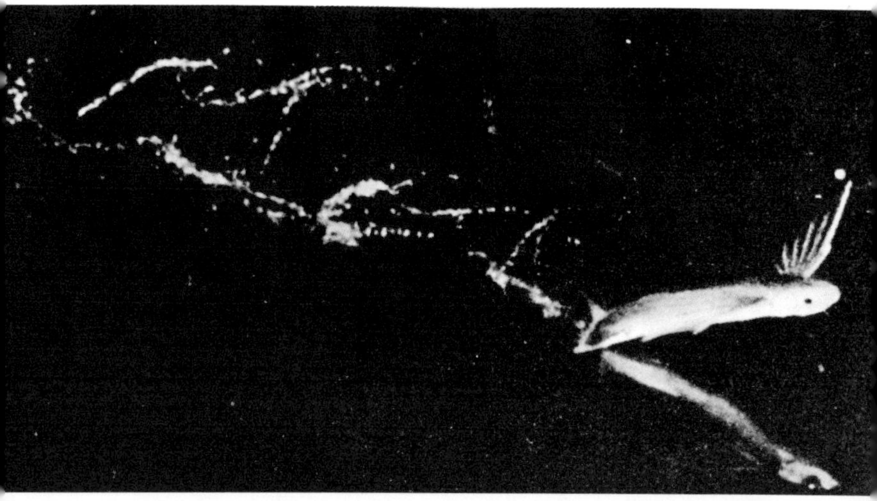

Not only prey fish school; the smaller predators also take advantage of schooling, and this for two reasons. First, they themselves are preyed on by the largest predators and thus need protection. Secondly, by attacking in schools they can fully surround a selected group of prey fish, thus preventing them from escaping, and by attacking from all sides they confuse the prey school, so that it breaks up and becomes more vulnerable. On the other hand, the climax predators (i.e. predators at the end of the food-chain) such as the giant tunny, do not school at all.

One fish deserving special mention is the tropical flying fish, which does not actually fly, but glides, using the enlarged pectoral and ventral fins as immobile wings. When this fish, probably fleeing from enemies, breaks the surface, it uses its tail fin to drive it forward and goes on using this vigorously at the surface until the body is completely airborne. Once in the air it can glide as far as a hundred metres, and so has a very good chance of avoiding its pursuers.

Plankton-eating sharks

The most curious fish inhabiting the surface waters are the Basking Shark (*Cetorhinus maximus*) and the whale shark (*Rhinodon*). These attain a length of 15 and 18 metres respectively – much larger than the notorious man eaters. They have hitherto been recorded as quite harmless to man, but in 1967 three attacks by whale sharks on fishing boats were reported from Mauritius. Despite its enormous size the Basking Shark lives on quite small planktonic organisms, chiefly copepods (*Calanus*). It swims at the surface, with its large triangular dorsal fin free of the water, and is often followed by a large flock of gulls, razor bills, cormorants, and gannets. The gill slits on both sides are so long that they almost meet both on the back and on the belly, as if the head were separated from the body. Each gill is supplied with an elaborate comb-shaped gill-raker, specially designed to sieve small animals from the water. With its immense mouth wide open, and with water streaming out through the sieves of the gill slits, this huge shark swims at a speed of four kilometres an hour. If it swims faster it pushes a column of water in front of it, making filtration less effective – much the same problem marine scientists encounter if a plankton net is hauled too quickly. It has been calculated that at its most economical speed the basking shark can filter over 2,000 cubic metres of water an hour, that is, more than 2,000 tonnes. The whale shark grows even larger, and can weigh up to a hundred tonnes. This animal also lives on plankton and uses much the same feeding technique as the basking shark.

Reptiles

Among other vertebrates in the nekton are the sea snakes of which we know fifteen genera and more than sixty species. They are unknown in the Atlantic, but abundant in the Indian Ocean, and most of them spend their entire lives in the open sea: They are able to swim backwards as well as forwards, and they live exclusively on fish. They are viviparous and just as venomous as the cobra, although sluggish and not very aggressive. Nevertheless, every year they kill several people, and a serum or a vaccine against their neurotoxic venom has, so far, not been manufactured. During half an hour's cruise in the Persian Gulf in 1937, the writer noted no less than twenty-nine sea-snakes sleeping at the surface, where they are especially abundant at dawn and sunset. The most common species seems to be *Pelamis* (*Pelamydrus*) *platurus*.

Sea turtles, some of which may attain a carapace length of one and a half metres and a weight of 500 kilograms or more, are typical members of the nekton. Usually restricted to the tropics, they spend most of their life in the surface waters of the open oceans. When the breeding season approaches, they migrate towards selected beaches

where they deposit their eggs in the hot, humid sand. The young turtles, usually hatching from the eggs at night, seem to have a certain 'compass sense' which leads them along the shortest possible distance to the shore. Adults provided with small radio transmitters fastened to the carapace have been traced on a migration of 2,200 kilometres from Ascension Island to the coast of Brazil.

The green turtle feeds on sea grass, while the loggerhead turtle and, largest of them all, the leathery turtle, are predators feeding mainly on fish. Leathery turtles are sometimes carried by the Gulf Stream as far as the coasts of Scandinavia. The populations of the green turtle, the so-called 'soup turtle', regarded as a special delicacy, have been exploited so hard by man during the last two centuries that speedy and strict protection is needed to avoid extermination of the species.

Mammals

Prominent among the members of the nekton are mammals that have forsaken the land on which they evolved and returned to the sea. These can be divided into two main groups – the many species of seals, which spend most of their lives in the sea but come ashore to breed, and the whales, which spend their entire lives at sea.

The whales are themselves divided into two families: the toothed whales (Odontoceti) range from the large sperm whale, to killer

3·3 A sketch based on a 16mm motion film, showing the arrival and departure of mature Atlantic sea turtles, *Lepidochelys kempi,* before and after laying their eggs in the sand dunes on the left. (After H. H. Hildebrand 1963.)

whales, dolphins and porpoises. The whalebone whales (Mysticeti) include the huge but now nearly extinct blue whale, as well as the fin, sei, humpback and several smaller species.

The toothed whales are predators, and gifted with a turn of speed which enables them to catch squids and large fish. A killer whale has been observed circling a ship moving at 38 kilometres an hour. The whalebone whales resemble the whale shark in that they all feed on plankton, and get their nourishment by sieving euphausiids from the surface waters by means of an elaborate strainer of baleen that hangs from the roof of the mouth. Thus the whalebone whales cruise steadily at low speeds, in contrast to the predatory toothed whales. Nevertheless, they are capable of high speeds (some 30 kilometres an hour) when on their biannual migrations between the summer feeding grounds and the winter breeding areas.

Marine mammals also comprise the sea cows (*Sirenia*), clumsy, walrus-like creatures which feed on sea grass. One group, the Manatees, inhabits both coasts of the tropical Atlantic, while the Halicorids are restricted to certain areas of the west Pacific.

Birds

The influence of birds on oceanic surface life is often disregarded, although untold millions of auks, guillemots and puffins populating the Arctic bird-rocks take all their food from the sea. Thus, a colony of some two thousand gulls (*Rissa tridactyla*) at Spitzbergen consumes about six million euphausiids a day. Off Peru, the 30–40 millions of guano-producing birds are known to eat some four million tons of anchovies every year.

4 Life and death in the photic zone

We have seen that the prerequisites for the production of vegetable matter (on which all other marine life depends) are light and nutrients. This means that the greatest biological activity will usually be found in the top 200–300 metres. This statement must however be qualified in two respects: first, light penetration varies widely according to the amount of turbidity, so that in extreme cases there is not enough illumination for photosynthesis to take place below fifteen metres. The second qualification is that although one would expect to find the greatest photosynthetic activity in the clear blue waters of the tropics, such waters are notably barren because of a scarcity of nutrients. The reason is that in the tropics the strong sunlight warms the surface layer, thereby reducing its density so that the nutrients remain at the surface. Furthermore, a discontinuity layer develops at between 100 and 200 metres, where the temperature and hence the density changes sharply within a few metres, which effectively prevents mixing of surface and deeper waters. In this situation the surface layers become depleted of nutrients and hence of plankton, which is why tropical waters, especially stagnant whirlpools like the Sargasso Sea, are very clear and intensely blue.

Not all tropical waters are barren, however. There are places such as the coast of Peru and Walvis Bay in south-west Africa where the prevailing wind blows offshore, driving the surface water out to sea; consequently colder water, containing a rich supply of nutrients, wells up from below and gives rise to high level of productivity. For comparison, the production of phytoplankton per unit area at Walvis Bay is eighty times greater than that of the Sargasso.

Phytoplankton and herbivores

Most of the micro-algae are consumed by grazers such as copepods and euphausiids. It is, at first sight, puzzling to find that, in the water column below a given area of the surface, the weight of herbivores (measured as organic carbon) is one and a half to two times the weight of phytoplankton. It seems as if the ratio of vegetation to grazer is upside down; but the explanation is that most of the phytoplankton is eaten almost as quickly as it is produced, and that a new generation can grow up in a week or less; the population of herbivores expands much more slowly.

This balance between phytoplankton and grazers holds good as a broad generalisation, but there are many occasions in which the relationship is more complicated. For instance, it has long been known that where micro-algae have been particularly dense there seem to be remarkably few animals. This was first explained by supposing that the mesh of a plankton net was so choked by swarms of algae that it could not sieve the water mass quickly enough to capture the more mobile animals. No doubt this is partly the reason, but it is far from explaining the whole problem. It is interesting, however, to note that relatively fast swimmers like the herring, able to switch with ease from bad to good food areas, avoid places with very dense algal swarms. In fact, many observations tend to show that the waste and metabolic products of algae can have a poisonous effect, or at least act as a deterrent to animals. The planktonic animals themselves also excrete waste products, and where such animals are present in dense shoals, these wastes may well be aggravating to larger animals such as fish. In recent years this problem has been seriously investigated by marine scientists, but it is still a long way from being understood.

Sagitta

Apart from the many plant-eating animals that practically dominate life in the illuminated zone, we also have a number of predators, the most interesting of which is the arrow worm *Sagitta*. This is a very old animal group. Fossil impressions have shown that these

4·1 The arrow worms (**a**) *Sagitta setosa* and (**b**) *Sagitta elegans* are indicators of poor and nutrient-rich North Sea water respectively. (After G. Thorson 1961.)

animals must have lived at least 500 million years ago. They are two to three centimetres long, slender, arrow-shaped, and transparent except for two small black eyes. The large throat is surrounded by curved bristles. The arrow worm usually floats without moving and is almost invisible, until with a sudden flick of the tail it pounces on a herring larva or a copepod, which is usually swallowed whole. The rate of digestion is high, the appetite voracious. And yet there would be no special reason for mentioning the arrow worm in preference to so many other predators from this zone but for another important characteristic – that of great discrimination in choosing its water mass. For this reason *Sagitta* species are 'indicators' and can provide marine scientists with important information about movements of large bodies of water.

Sir Frederick Russell, formerly director of the Marine Biological laboratory at Plymouth, made an important observation, namely that the surface layers off southern England can consist of two completely different water masses, the distribution of which can change from season to season and from year to year. These are oceanic water rich in nutriment, and North Sea water poor in nutriment. To separate these two water masses by chemical analysis is both difficult and time consuming, yet we know that their biological effects are widely different.

For example, Douglas Wilson, another able zoologist from the Plymouth laboratory, demonstrated that certain sea urchin larvae thrive excellently and develop normally in the rich ocean water but develop abnormally or even perish in the poor North Sea water. Now it is known that each of these two water masses has its own absolutely characteristic arrow worm. *Sagitta elegans* is an indicator for the oceanic water; *Sagitta setosa* for North Sea water. In order to avoid lengthy and laborious chemical analyses one need capture only a couple of arrow worms and identify them to determinate the kind of water being investigated. It is easy to distinguish between

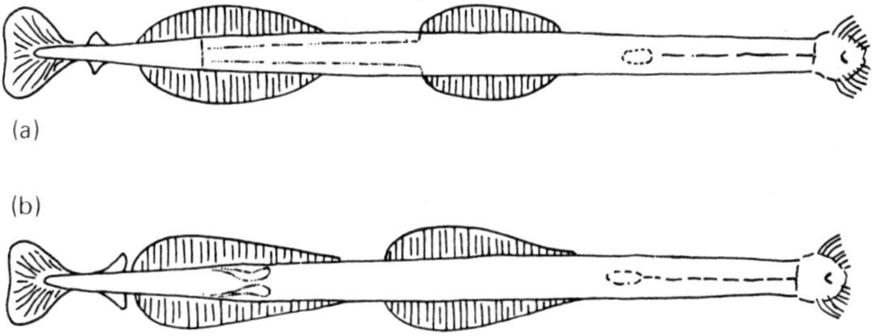

the two arrow worms, first by the shape and arrangement of the fins, and secondly by preservation in formalin, when *elegans* becomes opaque while *setosa* remains transparent. As the two water masses have a great biological influence on the areas they cover, Russell's discovery has wide practical importance. For example, in the North Sea there is also an obvious difference between the bottom fauna, dependent on whether this is found in an area characterised by *setosa* water or *elegans* water. Incidentally, it has more recently been found that a definite colour difference can be seen in the two water masses in the Plymouth area when observed from an aircraft.

Russell's discovery of indicator species has led to similar and equally successful observations elsewhere. Not only have characteristic species of arrow worms been found for a large number of different water masses, but other groups of planktonic animals – copepods, pteropods, ostracods, and certain tropical shrimps – have proved to be important indicator species. In this way, zoologists have given hydrologists a simple and practical method for a quick discrimination between water masses.

Larvae and mortality

One extremely important component of the plankton of the photic zone consists of the larvae of bottom-dwelling animals. It is calculated that there are about 135,000 species of such invertebrates and that more than 100,000 of these have a free-swimming larval stage lasting for two to four weeks. The bottom dwellers that produce

4·2 Free-swimming larval gastropods (veligers) from Danish plankton. (**a**) *Philbertia linearis*; (**b**) the periwinkle *Littorina littorea*; (**c**) *Mangelia attenuata*; (**d**) *Bittium reticulatum*; (**e**) *Nassarius incrassatus*; (**f**) *Balcis distorta*; (**g**) *Nassarius reticulatus*; and (**h**) *Limapontia capitata*. (After G. Thorson 1968.)

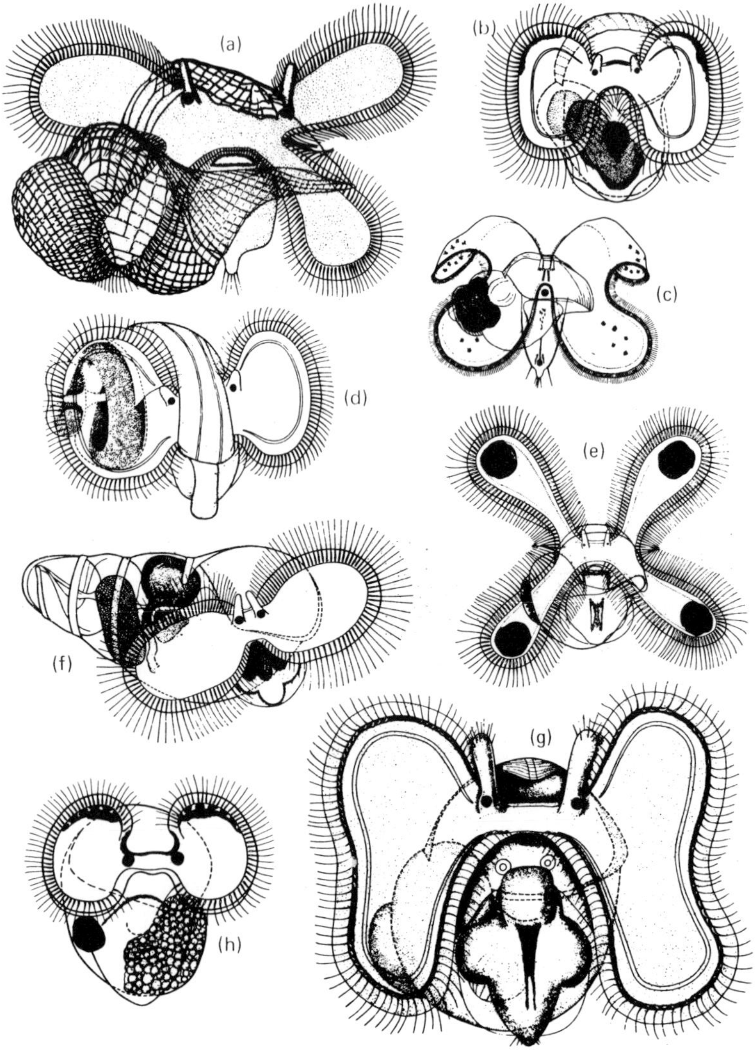

these larvae are themselves confined to relatively shallow waters of the continental shelf. Their larvae are thus present only in coastal waters, and indeed predominate in the plankton from time to time.

Such larvae occur in astonishing numbers, simply because of the huge egg production of the adults. A rough calculation for twenty-six different species of bristle worms, snails, mussels, echinoderms, and crustaceans shows an average production per female per year of 60 million eggs for species in which the larvae have a long pelagic life and find their own food. The enormous number of eggs immediately tells us that mortality during development must be very great. All that is required of the parents – when a female spawns, and when a male fertilises this vast number of eggs – is that they should maintain the population, that is to say that all the eggs laid need ultimately result only in one sexually mature female and one sexually mature male to replace the parents that produced them. If there were to be a surplus of larvae reaching sexual maturity over a number of generations the species in question would quickly achieve almost total dominance on the bottom. If, on the other hand, a series of generations produced a deficit in propagation intensity, the species would soon become extinct. However it appears that, in waters which have been under investigation for the last hundred years, the balance between the total number of bottom animals among the different species is, on the whole, fairly constant, and that the ratio of the number of eggs produced and the loss during development also remains constant.

These considerations show that the more eggs spawned by a species the greater is the larval mortality. Forms with large eggs rich in yolk which develop into young bottom stages without a free-swimming stage can often manage with a few hundred eggs per year; by contrast, it is evident that the greatest loss by far takes place while the larvae are swimming freely in the surface layers.

What are the reasons for these enormous losses during the free

4·3 Examples showing viviparity, brood-protection and other types of non-pelagic development in Arctic bottom invertebrates. (**a–c**) Sea anemones; (**d–f**) sea stars; (**g**) a scale worm; (**h**) a sea cucumber; (**i**) a bivalve with egg strings; (**j**) prosobranch egg capsules; (**k, l**) limpets with young in egg cavity; (**m**) a large shrimp with giant eggs clinging to its abdomen. (After G. Thorson 1951.)

larval life? Many bottom animals spawn their eggs and sperm in cloud-like forms which quickly disperse, and it seemed that the problem was to ensure that a sperm cell and an egg of the same species had a reasonable chance of meeting in those vast water masses during the few hours in which they were able to fertilise each other. For many years zoologists believed that herein lay an important source of loss, and that a large number of eggs disintegrated without having been fertilised. We now know, however, that in many bottom-living animals the male spawns first, while the female waits to spawn until she is surrounded by a sperm cloud of her own species. When the eggs are directly spawned into a water mass with sperm of the same species the chance of a large percentage of fertilisation should be considerable. In other cases it is known that the male does not spawn first, but the whole population, males and females, spawns simultaneously. This seems to show that high mortality is not, to any extent, due to failure of fertilisation.

Another observation supporting this view is that certain copulating snails such as the sea hare, *Aplysia,* in which practically all eggs are fertilised before being spawned, produce as great a number of eggs as that produced by species which shed their sperm and eggs freely in the sea. In both cases the heavy wastage must, therefore, take place after spawning, during the pelagic larval phase.

The reason for heavy losses must therefore be sought at a later stage, during the development of the larvae in the plankton. We have two main types of pelagic bottom-living animal larvae: the first originates from large, yolk-rich eggs which contain so much nourishment that the larva is not obliged to seek outside food during the free-swimming stage. The second larval type originates from small eggs, poor in yolk, and during its free-swimming life it is completely dependent on the algal food it must itself obtain. A good example of the contrast between these larval types is to be seen in two Scandinavian species of sea stars, the adults of which are much the

(a)

Epiactis marsupialis Carlgr.

(b)
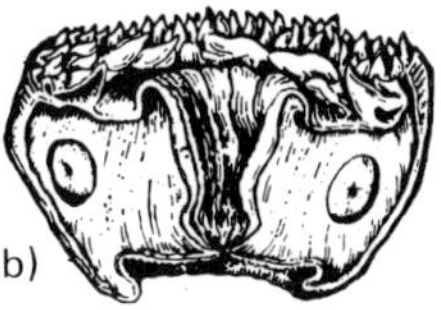
Actinostola sibirica Carlgr.

(c)

Epiactis prolifera Verr.

(d)

Leptasterias groenlandica (Stp.)

(g)

Harmothoë imbricata L.

(e)
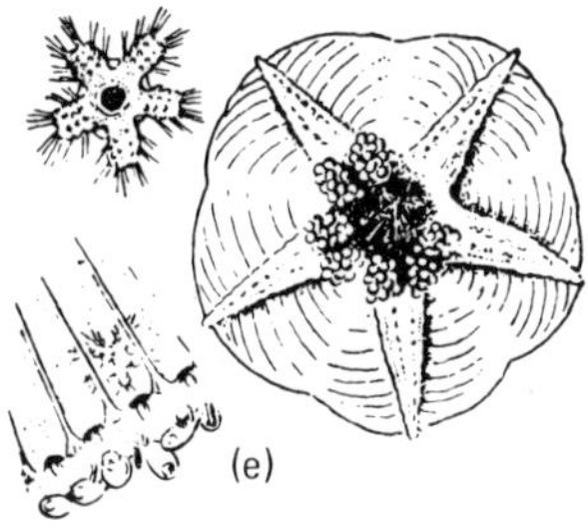
Hymenaster pellucidus W.Th.

(h)
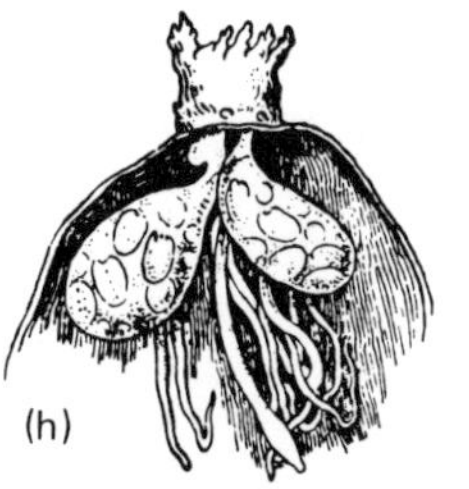
Cucumaria glacialis (Ljungm.)

(f)
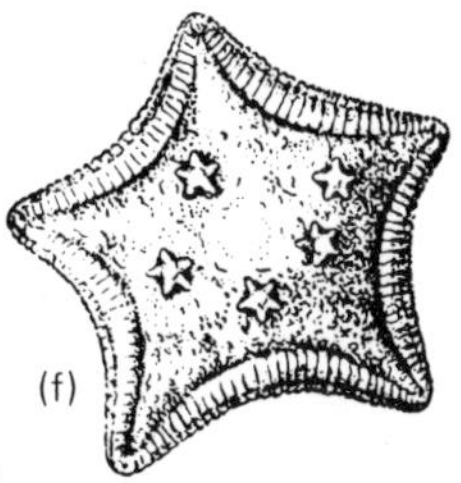
Trophodiscus almus Fish.

(i)
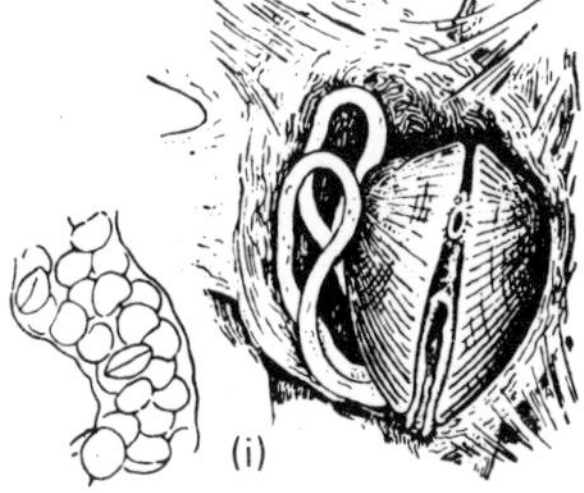
Modiolaria laevigata Gray

(j)
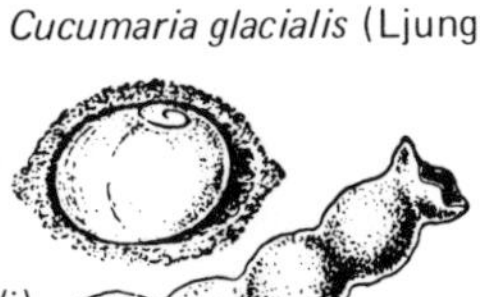
Acrybia flava (Gould)

(k)
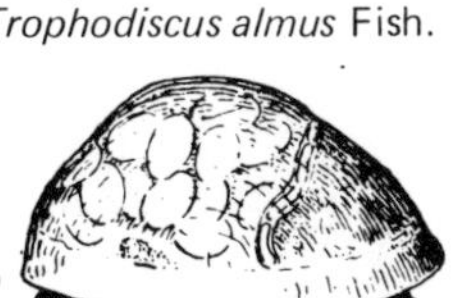
Acmaea rubella (Fabr.)

(l)

Volutopsis norvegica (Chemn.)

(m)
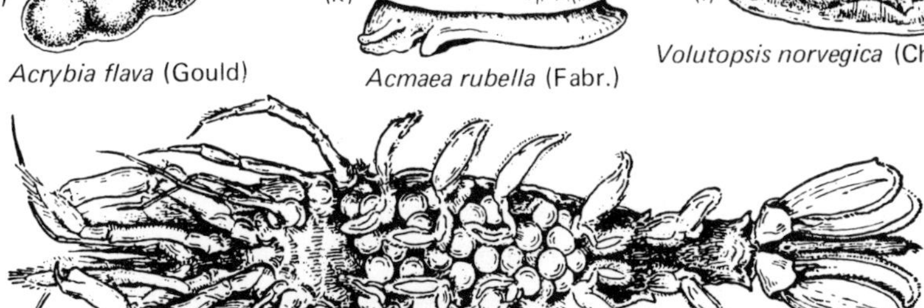
Sclerocrangon ferox G.O. Sars

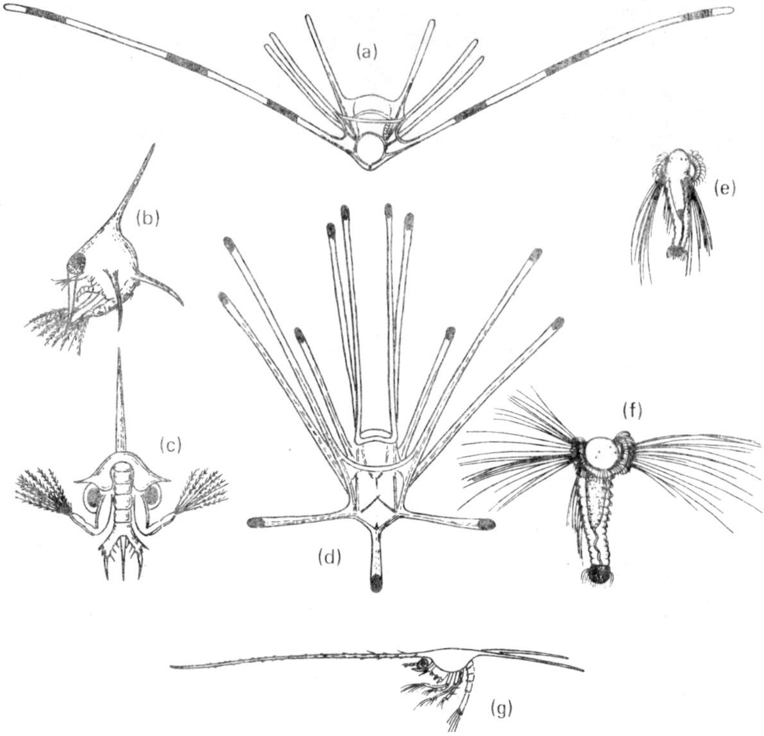

same size. The first species produces yolk-rich eggs and hence larvae that do not need to seek food during the pelagic stage; the second species has eggs that are poor in nutritive value and hence larvae which must feed themselves during their pelagic phase. The first species can survive with 10,000–20,000 eggs per season, while the other must produce 2,500,000 eggs within the same period. In other words, the greatest mortality takes place in those larvae dependent for food on algae or other plankton.

Lack of food, then, appears to be one of the reasons for this loss. Nevertheless, if one goes into details it is evident that most larvae are well able to tolerate starvation conditions; but hunger impedes development and prolongs the period spent in the free-water masses. Another contributory cause of mortality is a sudden temperature

4·4 Larval bottom invertebrates with appendages which make them difficult for predators to swallow. The larvae belong to (**a**) a brittle star; (**b**) a pea crab; (**c**) a swimming crab; (**d**) a heart urchin; (**e**) and (**f**) polychaete worms; and (**g**) the decapod *Porcellana*. (After G. Thorson 1968.)

variation, which can give larvae a temperature shock. This seldom happens in the sea, but it can happen that the temperature throughout the entire sea area remains for a period below the normal, during a cold summer for instance. This will not actually kill the larvae but it will reduce their metabolism and – just as with lack of nourishment – delay their development, thus retarding their descent to the bottom.

When the time comes for returning to the bottom the larvae are exposed to yet another crisis. Some live on clay bottom, others on sand, others on rocks or reefs: each must find its own particular habitat. Again, it was long assumed that during this process there was a heavy mortality, a theory based on the assumption that on reaching the descent-to-bottom stage, the larvae scattered in a haphazard manner, and that it was purely a matter of chance which type of bottom they encountered. We now know that this interpretation is wrong, and again it is Douglas Wilson of Plymouth who has put us on the right track. Many groups of larvae migrate, as soon as they are hatched, towards the light and hence the surface. Towards the end of their pelagic phase they shun the light and seek the bottom. There then follows a swimming-crawling stage, in which the larvae penetrate into the bottom material and investigate it. If they find it suitable they can remain, turning into young bottom stages in the course of a few hours. If, on the other hand, the substrate does not suit them they can resume the free-swimming life. By continually shunning the light they are able to prolong their larval life by days or even weeks, while the current carries them slowly along the sea bed until they find their correct environment. In this way, far more larvae make contact with the substrate suitable for their needs than was formerly believed possible.

Nevertheless, it is a fact that an overwhelming number of larvae are lost during the free-swimming stage. What, then, is the reason for this? There is one thing that lack of food, low temperature, and

4·5 Below Many different predators eat their share of the free-swimming larval bottom invertebrates. The photos show four predators which were offered oyster larvae as food in an experimental tank. (Top left) Cirri of adult barnacle with oyster larvae (black dots) sticking to them; (top right) larvae filling out the whole digestive cavity of the polyp of *Aurelia*; (bottom left) larvae in the gut of the bristle worm *Polydora*; and (bottom right) in the gut of *Sagitta*. (After Cerruti 1942.)

4·6 Right The hydroid *Hydractinia echinata* living on the shell of a hermit crab, shown from the moment it (**a**) captures a copepod; (**b**) swallows it; (**c**) regurgitates the copepod cuticula; and (**d**) twenty-six hours after capture, when it is ready to tackle a new prey. (After G. Thorson 1965.)

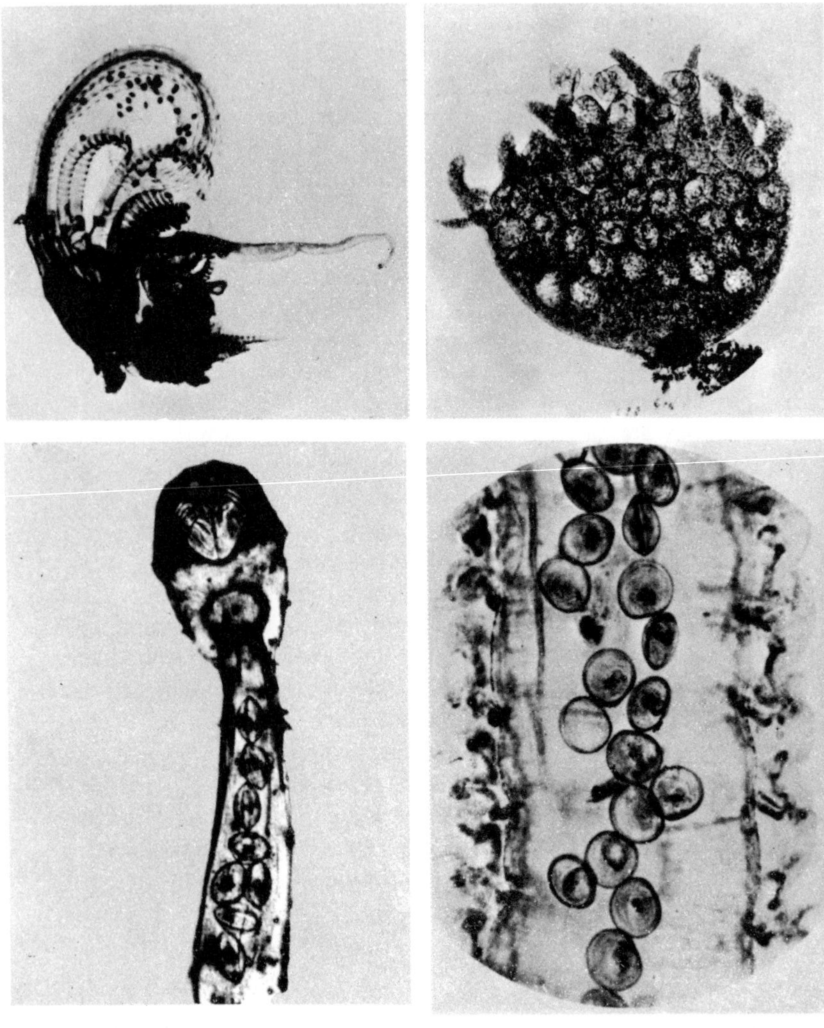

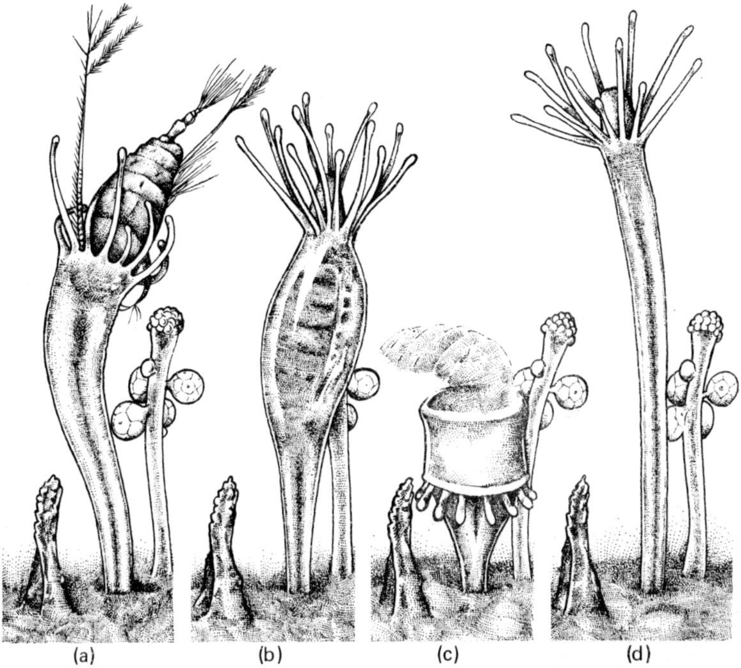

unsuitable substrate have in common: they all lead to a prolongation of the free-swimming larval life. Each extra day in the free water masses means twenty-four hours of merciless attack from large numbers of enemies, against larvae which are completely unprotected and have very little chance of escape. It has been calculated that about 10 per cent of a larval oyster swarm is eaten every day. As has earlier been mentioned, even unicellular organisms such as the large flagellate *Noctiluca* eat considerable quantities of these larvae. Also, the larvae are not too considerate of each other. The large larvae of some polychaete worms have bivalve larvae as their favourite food; crab larvae eat larval molluscs on a large scale and are themselves eaten by lobster larvae, and so on.

There are also other groups of voracious predators, such as the medusae of the hydrozoans, the comb jellies and the arrow worms. One species of comb jelly, *Mnemiopsis leidyi,* which lives in west Atlantic waters, is reckoned to swallow 472 copepods of the genus

Acartia every hour. A count of the average density of *Mnemiopsis* means that these comb jellies will kill 22,000 copepods per cubic metre of water every twenty-four hours; these comb jellies are also known to feed, on a large scale, on oyster larvae. Larval fish, too, especially the young stages of herring larvae, are known to eat large quantities of bottom animal larvae.

When the larvae reach the stage at which they metamorphose and settle on the bottom, yet another group of enemies is waiting for them. Even so seemingly innocent an animal as the common mussel, which lives by filtering micro-algae out of the water, contributes to the slaughter of larvae, including larvae of its own species. A medium sized mussel filters about 1·4 litres of water an hour, and all larvae in this water mass are trapped in the meshes of the mussel gills, embedded in mucus and, as a rule, expelled as 'pseudo-faeces'. During the season, when the larval swarms are at their maximum in a shallow Danish fjord, a mussel can trap and kill about 100,000 bottom animal larvae in twenty-four hours.

There are, however, also many true enemies on the bottom – predators sitting and waiting for the larval swarms. Examples are the polyps of the hydrozoans, among which the polyp-generation of the common jelly fish *Aurelia* is especially voracious. Even sea anemones, barnacles, brittle stars and newly settled sea stars are ready to take their large share of larvae when these approach the bottom.

To sum up, the chief reason for the enormous loss among bottom animal larvae is that they fall prey to predators; and perhaps we should mention here, once and for all, that to be swallowed by another creature is the natural destiny of well over 90 per cent of all sea animals.

5 A world without light

The twilight zone

It was mentioned earlier that at 200 metres depth only 1 per cent of even the blue light rays survive in the clearest water. At this depth the absolute limit for algal production has been reached; but we have seen that a fraction of the surface light goes even deeper and that even at 800 to 950 metres animals performing daily migrations up and down in the water appear to be influenced by the light.

This zone, which we call the twilight zone (the mesopelagial), with an upper boundary of about 200 metres and a lower boundary of about 1,000 metres, has a very rich animal life. The vast schools of small luminiferous fish of the genus *Cyclothone* deserve special mention. In this layer they are found over enormous areas of the oceans. One species, *Cyclothone signata,* is believed to be the commonest fish in the sea, and that is saying a great deal when one remembers how large a school of ordinary herring can be. Another group of luminiferous fish, the Myctophidea, is extremely numerous in this layer, which also houses several species of *Argyropelecus* – compressed fish with upturned mouths, telescopic eyes and large luminous silvery plates along the length of their body. *Stomias* should also be mentioned here – long, slender fish with enormous mouths, large teeth pointing backwards, long, whisker-like excrescences from the lower jaw and light organs running along the sides like port holes on an ocean liner. In addition, there are huge swarms of squids, euphausiids and copepods.

At first it may seem strange that this particular part of the free water masses should have such a rich and varied animal life, because plant food is produced only in the surface layers. But in actual fact, life in this zone offers several important advantages. In general, the illuminated zone has much warmer water than the deeper lying twilight zone. As the metabolism, and hence the food demand, decreases with decreasing temperature, animals living in the twilight

5·1 Plankton and nekton from the northern Pacific, between 280 and 850 metres (twilight zone). In the uppermost layer are the amphipods *Parathemisto gracillipes* (large eyes) and *Cyphocaris challengeri;* two *Euphausia pacifica;* a few copepods; and (in the background) two siphonophores, *Nanomia bijuga.* In the centre are two fishes – the larger *Lampanyctus leucopssarus* and the smaller *Diaphus theta.* Near the bottom of the picture are two cephalopods, *Loligo peali;* the shrimp *Sergestes similis* (very long antennae); and the amphipod *Onchomenella nana.* (Kai Olsen.)

zone are able to manage with much less food than those living nearer the surface. Together with this advantage, and the much more stable temperature, they are still close to the food-producing layer and are the first to benefit from the continuous fall of nourishment (both plants and dying or dead animals) which sinks towards the bottom. Moreover, we have seen how the viscosity of cold water is much greater than that of warm water, making it easier for animals to float. When plants and animals adapted to floating in the warm surface water sink down into the colder water, they encounter an increase in friction, which either stops them completely or makes them sink very slowly. The result is that the twilight layer serves as a storehouse for sinking organisms, and the number of predators, which probably includes carrion eaters, is accordingly much larger than in the upper layers.

All the same, the animals of the twilight zone do not simply wait for food to float down to them. Many are known to undertake long diurnal migrations, ascending when dusk falls and descending at daybreak. Most of these up and down migrations are in the region of 100–200 metres, with the animals going up into or close to the food producing surface layer, but there are cases of similar migrations of between 1,000 and 2,000 metres. Here the animals never actually reach the algal producing layer, but nevertheless they undertake these long distances twice every twenty-four hours. Vertical migrations of this type are known to take place among several different animal groups: medusae, jelly fish, ctenophores, arrow worms, pteropods, many groups of crustaceans, holothurians, and young and adult fish. At one time such migrations were thought implausible. As one catches far more species and specimens with a net in the upper waters at night than during the day, it could be (and was) argued that the fastest animals would see the nets at a

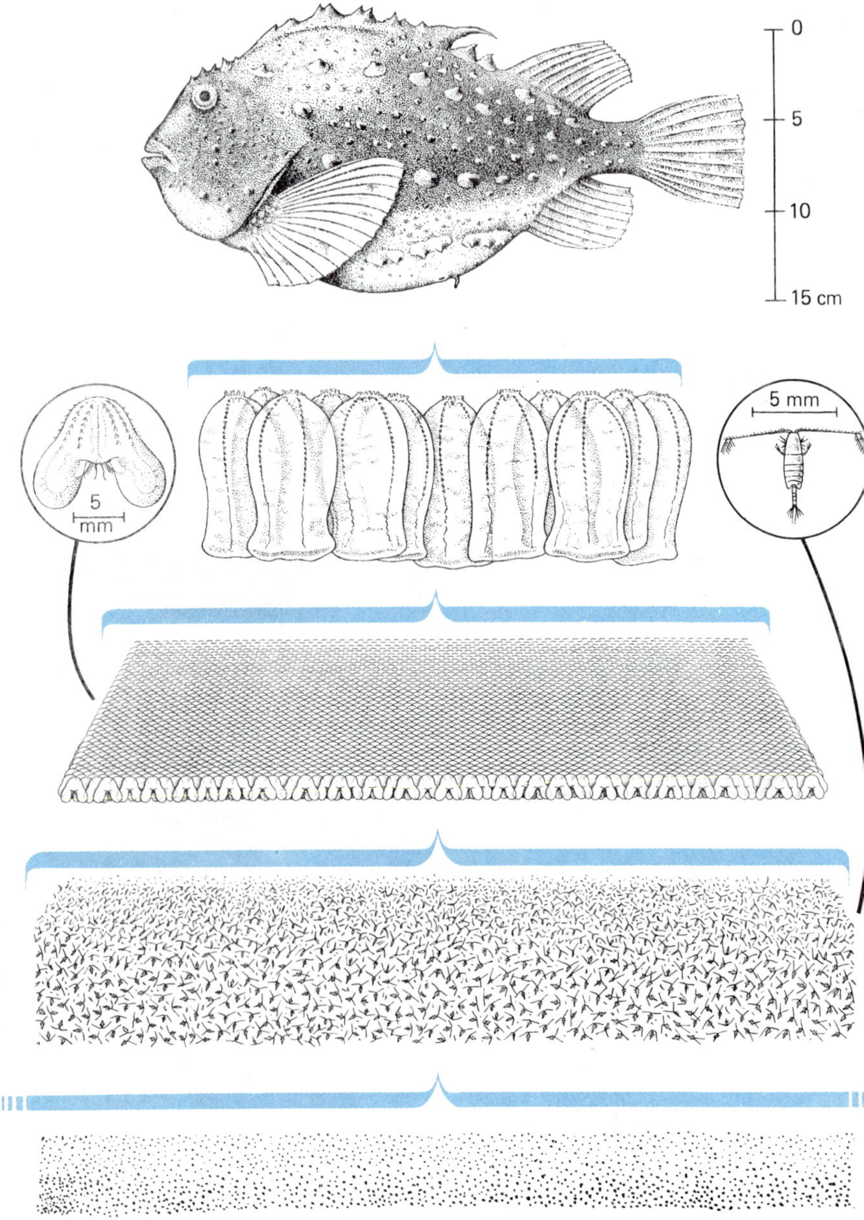
0
5
10
15 cm
5
mm
5 mm

5·2 Diagram illustrating a food chain in the plankton. A lump sucker (*Cyclopterus*) (top) will swallow twelve large comb jellies (*Beroë*) in one meal. To reach the size they are when eaten, these *Beroë* will probably have eaten a total of some 3,500 specimens of the smaller comb jelly (*Bolinopsis*) (circle, left). The *Bolinopsis* in their turn will have swallowed a total of some 100,000 copepods (*Calanus*) (circle, right), which again will have devoured untold millions of diatoms. (Kai Olsen.)

sufficient distance to avoid them during the day, but were taken by surprise in the dark. We now know that this argument does not carry any weight; thorough investigations have shown both that there are more species in the surface layers and fewer animals in the deeper layers, at night.

It has been shown experimentally that copepods can swim upwards at a speed of 17–30 metres an hour, and the same is true of some crustacean larvae. The 2–3 centimetre-long shrimp-like euphausiids can swim upwards at a speed of 100 metres an hour. The descent is, of course, quicker: 50 metres an hour for copepods and about 140 an hour for euphausiids. These speeds agree with vertical migrations that have been observed under natural conditions. The ascent in particular costs a great deal of energy, and since animals living a highly competitive life can hardly spare energy on fruitless activity these migrations must have some significance and offer some advantage.

It is, of course, an advantage for the animals to place themselves as close as possible to the food source, and ideally to penetrate the food-producing layer itself. But then one must ask why they do not remain there. One possible reason why so many different groups take part in these migrations could have been that the herbivores go up after algae, and that they are in turn followed by predators which live on plant-eaters. In large fresh-water lakes similar up and down migrations take place among the plankton, and observations made on the small water flea (*Daphnia*) would seem at first glance to support this explanation. *Daphnia* becomes appreciably heavier after a meal, and if we apply this observation to marine plankton it might be supposed that animals migrate up towards the food layer during the evening, eat algae at night, become heavier, and automatically sink back again. But this explanation will not do, first of all because it does not explain why all animals should ascend, eat,

5·3 The distribution of the copepod *Calanus finmarchicus* in the surface water masses during a twenty-four hour period from noon to noon. (After F. S. Russell 1927.)

and again sink simultaneously. Secondly, the weight difference found before and after a meal is not enough to account for such large migrations, and thirdly, experiments have shown clearly that descending animals not only sink because of their weight but actively swim downwards. Finally, the idea that food is the chief reason for migrating is countered by the fact that similar daily migrations (from 200 metres to 1,000 metres and back) are undertaken by a number of species which never reach the food layer.

One could, perhaps, argue that as all these animal groups seek the surface only in the dark, it is because they are invisible at night and can better avoid enemies. But many of the migrants, euphausiids for instance, are luminiferous and very easy for predators to spot.

But there is one more factor which may play a role in these migrations – the light. Almost all plankton animals have marked reactions to light: many are photopositive, that is they are attracted towards the light; others are photonegative – they shun the light. However, the matter does not end here: plankton animals are very sensitive and can easily switch over to the opposite reaction. For example, many plankton animals are strongly attracted by light at a certain strength, but if the strength is doubled they will react in the opposite way and try to avoid the light. This change from positive to negative occurs more quickly in waters of high temperature and decreased salt content. And as a further illustration of how complicated these reactions can be, one may use the above mentioned *Daphnia* as an example: these animals (which, as we have said, become heavier when they have eaten) have a tendency to seek the light when hungry and avoid it when they have eaten their fill.

It would appear, therefore, that light plays an important part in the vertical migrations of planktonic animals; and the most acceptable and simplest explanation has been given by Sir Frederick Russell at Plymouth. After making an impressive collection of planktonic animals from around southern England, using a closing net at

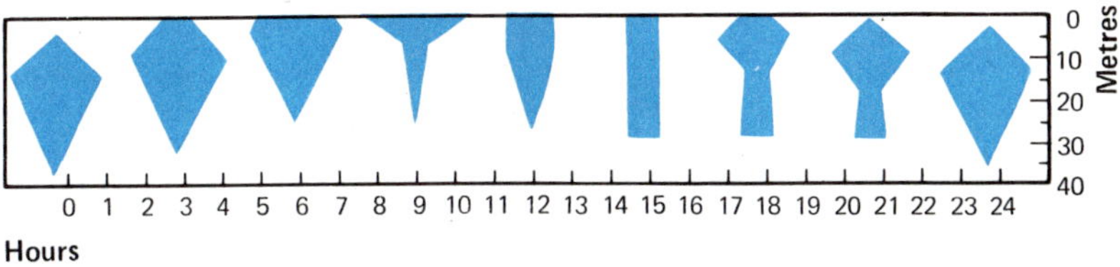

different depths and at all times of the day and night, Russell concluded that the animals of the twilight zone lived in this layer because, among other things, they preferred a rather weak light to complete darkness. During the day they found this pale light in the twilight zone, but towards sunset they followed the diminishing light up towards the surface, reaching it early in the evening. Around midnight, when even the surface was dark, they dispersed again; but with the first morning light they were once more attracted by the weak-light zone, and followed it down as the sunlight increased in strength and penetration.

Russell has no doubt found an important link in the 'mechanics' of migrations, but even so the whole process is not fully clear. There is no doubt about the importance of light but there are still discrepancies. For instance, copepods react in different ways to light in different generations, and now and again one comes across specimens which normally live in the depths but are found swimming in the clear sunshine of the surface. Other factors must be involved in the process, but even if these were known they would provide only a purely technical explanation, revealing nothing about the advantages the animals gain from such energy-consuming migrations.

The English zoologist, Sir Alister Hardy, who has been particularly interested in this question for many years, points out yet another factor: by undertaking the vertical migrations, and in no other way, animals are able to make a radical change in habitat. Swimming horizontally, planktonic animals cannot travel far. But the current is nearly always fastest at the surface – often between one and two knots faster than at a couple of hundred metres depth. By swimming towards the surface at night the animals obtain free transport, and when at daybreak they sink again they find themselves in a completely different water mass.

Vertical migrations seem, furthermore, to be important mechanisms for the transport of food from the surface layers to the deep

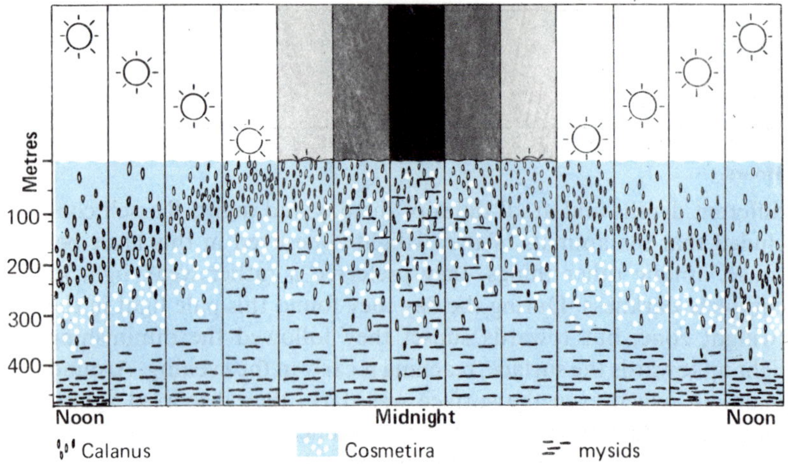

sea. The Russian biologist M.E. Vinogradov pointed out that if an organism dies in the surface layers it starts to sink, but only one-tenth to one-thirtieth of the particulate organic matter produced in the upper layers reaches 4,000 to 5,000 metres and this usually consists of indigestible parts of skeletons. At depths of 6,000 to 6,500 metres in the Kurile-Kamchatka Trench we find a thousand times less plankton than in the topmost fifty metres. Animals descending from their nightly feeding migrations into the surface layers will, at a few hundred metres, meet the ascending inhabitants of deeper waters; these feed on them and on their remains and these too, after descending to still greater depths, become, in their turn the prey of animals living in the deepest layers. Thus it follows that the main mass of deep-sea plankton consists of predators.

The scattering layer One of the byproducts of the Second World War has been a great increase in our knowledge of vertical migrations. In the fight against U-boats echo conditions in the water masses were very important, and the nations at war were naturally anxious to improve their echo-sounding methods. An echo sounder works by transmitting a sound pulse of brief duration to the bottom; the echo returns to the surface and is picked up by the ship's microphones. Since we know the velocity of sound in water, by measuring

5·4 The diurnal vertical migration of three common species in the plankton: copepods (*Calanus*); jelly fish (*Cosmetira*); and fairy shrimps (mysids). (After J. Fraser 1962.)

the time lapse between the sending of the pulse and the reception of the echo, we can calculate the depth with great accuracy.

During trials with this equipment in the deep waters of both the Pacific and the Atlantic, the American Navy often found that as well as the desired echo from the sea bottom, another echo was received, originating from somewhere in the water mass itself. At times this echo was so strong that it almost drowned the true echo from the sea bottom. The layer became known as the 'scattering layer', the 'ghost layer' or 'the false bottom', and was at first believed to be a purely physical phenomenon. Physicists could get nowhere with the problem until their material was seen by the Californian plankton expert, Martin W. Johnson; he was at once fired with enthusiasm when told that the echo came from 500–900 metres or even deeper during the day, but at night from only 100–200 metres depth. For a plankton expert it was an easy step to combine this information with our knowledge of plankton animal migrations, and thus the first part of the mystery was solved.

The next point, which made this discovery even more significant, was that these pelagic organisms were dense enough to give a uniform echo over a large area, and that one could register the echo of the 'false bottom' almost continuously throughout the depths of the Pacific, from the Arctic Aleutians to the tropical island of Tahiti and from San Francisco to Japan. Further investigations soon produced a similar echo layer in the Mediterranean, the Red Sea, the Indian Ocean and the tropical Atlantic. Thus, in these immense and hitherto unexploited areas, and indeed almost everywhere in the free water masses, there existed a layer of animals in such numbers that a strong echo was produced. These animals, it was quickly realised, could be utilised for human consumption.

While the war was in progress the US Navy classified its knowledge of the 'false bottom' as top secret. After the war, however, biologists inherited the problem and serious attempts were made to discover which organisms were responsible for the echo. On the basis of animal life already known to inhabit the 'twilight zone', plankton experts listed the following possibilities: euphausiids, squids, fish, or a mixture of them all.

Euphausiids appeared to be a very real possibility as these are often found in dense shoals in the water masses. The American planktologist Dr Hilary Moore calculated that in the echo layer off the coast of Florida there are about twenty euphausiids in every cubic metre of water – and the sea round Florida is not even particularly rich in these animals. New data from the Pacific coast of Canada have shown that the scattering layer here may contain up to 175 euphausiids per cubic metre. As we have mentioned, euphausiids are far more numerous in the polar seas than anywhere else, but thorough investigations in recent years have shown that there is an especially poor echo in the water masses of the Antarctic. Thus, although there might be large numbers of euphausiids in the echo layer, they can hardly be the cause of the echo.

Squids can occur in large numbers, as they do off the Pacific coast of the United States. Mostly small animals, they can be very numerous, both in the surface layer and below. However, these animals have a tendency to remain in rather dense schools, and it is difficult to imagine that they could form a layer uniform enough to create a scattering layer over vast areas of the open oceans.

Fish are the third possibility, and the most recent investigations seem to increase the likelihood that they are responsible for the echo. It is known that the air-filled swim-bladder of a fish reflects sound waves well. Experiments undertaken with true fishes and also with fish models containing an artificial air-bladder have demonstrated very distinct echoes, and it seems that even a relatively small number of fish dispersed in the water can generate a surprisingly big echo. Thus comparatively few fish, distributed in a certain water layer, may be responsible for the 'scattering layer', and the echo layer does not necessarily mean that such immense, unexploited sources of food exist as was thought until a few years ago. Nevertheless, an observation from the submersible *Alvin* in 1968 during a cruise off Washington at 1,800 metres, showed that a scattering layer consisted of schools of the fish *Ceratoscopelus*. The schools were five to ten metres thick and ten to one hundred metres in diameter. The centres of each school were a hundred to two hundred metres distant from each other and the density of fish was ten to fifteen per cubic metre.

It has recently been shown that coelenterates such as the siphonophore *Nanomia,* which has gas bubbles in its body, may also be responsible for a portion of the echo. The echo layer must, of necessity, comprise other and considerably more animal species than just a few scattered fish. The layer appears to be fairly constant, and the echo-producing animals undertake regular twenty-four-hour migrations up and down in the water column. The fish which supposedly supply most of the echo must, therefore, obtain nourishment in this water layer. These fish are, of course, predators and we can calculate that they, like other fish of this type, can utilise approximately 10 per cent of the weight of their prey to increase their own weight. Thus the animal food, which *must* consist chiefly of copepods and euphausiids, must also constitute a much larger total weight and volume than that of the predator fish. Even if the fish which produce the echo prove to be dispersed in the water mass, there must be a very considerable amount of prey in the same water layer for the fish to maintain life.

The British zoologists Cushing and Richardson have found that the dry weight of the organism in the deep scattering layer would be 5–10 grams per square metre, and according to still newer data at least some parts of the deep scattering layer may contain as much as 44 grams dry weight per square metre. All in all, the 'false bottom' still deserves close investigation from a food-producing point of view; no doubt it will sooner or later be included in man's plans for a more rational utilisation of the resources of the sea.

Bioluminescence The twilight zone also contains the largest proportion of bioluminescent animal species, especially shrimps, squids and fish. About four-fifths of all pelagic fish species in the depths around 500 metres carry light organs. The light they emit is, in most cases, formed in special light-producing photophores which, taking only one species of fish, may be organised in up to twelve different patterns on the body. A photophore usually consists of glandular cells or of glands containing a culture of luminescent bacteria surrounded by a cup of dark pigment. In particularly efficient luminous organs the radiation may be amplified by front-lenses.

The chemical process by which the light is produced consists of an interaction of the compound luciferin with the enzyme luciferase. It is a pronounced 'cold light' of the same kind that is produced on land by fire flies. This 'cold light' is the most efficient type of light production known so far. The fact that the light is 'cold' means that it is produced with only a minimal loss of calories. When, for example, we produce light to illuminate our towns, at least 50 per cent of the original energy is lost as heat, but a glow worm or a small crustacean can produce light with a heat loss of only 1 per cent.

Nevertheless, we see that these animals economise with light, which in most cases flashes on only when triggered by a nerve impulse from the animal itself, or from a stimulus from outside. It has been observed, for instance, that when a ship switches on its radar, animals in the surrounding water begin to emit light. Furthermore, it has been found that intermittent flashes are perceived much better by deep-sea animals than is permanent illumination; thus economy and practice act together in an exemplary way.

A large number of the bioluminescent species, and above all the fishes of this zone, carry their photophores only or mainly on the ventral half of the body, not on the dorsal side. Thus by daylight, and seen from below, the self-illuminated ventral side might help them not to appear dark against the background and so help to conceal them from predators. By night this may be of no help to them, but here the ventral position of the photophores may serve another purpose. Since the number of predators is greatest in the surface layers and decreases rapidly with increasing depth, there will usually be fewer predators below the prey fishes than above, and thus there is a greater possibility that a fish would be seen and eaten by predators if its bioluminescence were dorsal rather than ventral.

The deep sea zone

Below the twilight zone, the lower limit of which we defined as approximately one thousand metres, lies a huge mass of water which is called the Deep-Sea Zone, or Bathypelagial, accounting for 90 per cent of the living space of the oceans. Here there is perpetual darkness; here too there is a free connection between all the cold-water masses of the oceans (with the exception of the water north of the North Atlantic Ridge).

This has to a large extent influenced the distribution of animal life, which consists of extremely widespread, often cosmopolitan forms. In no other area of the globe is such a specialised life spread over such immense areas, and it is remarkable that this teeming life was unknown to man even a hundred years ago.

When in the open sea we move from the highest, illuminated zone, through the twilight zone, down to the deep-sea depths we encounter all the stages of a sort of 'inverted food pyramid'. In the top layer we have algae, plant eaters and predators. Hardly any plants are still alive by the time they sink into the twilight zone, and the remaining plant eaters mostly eat dead particles and organic detritus.

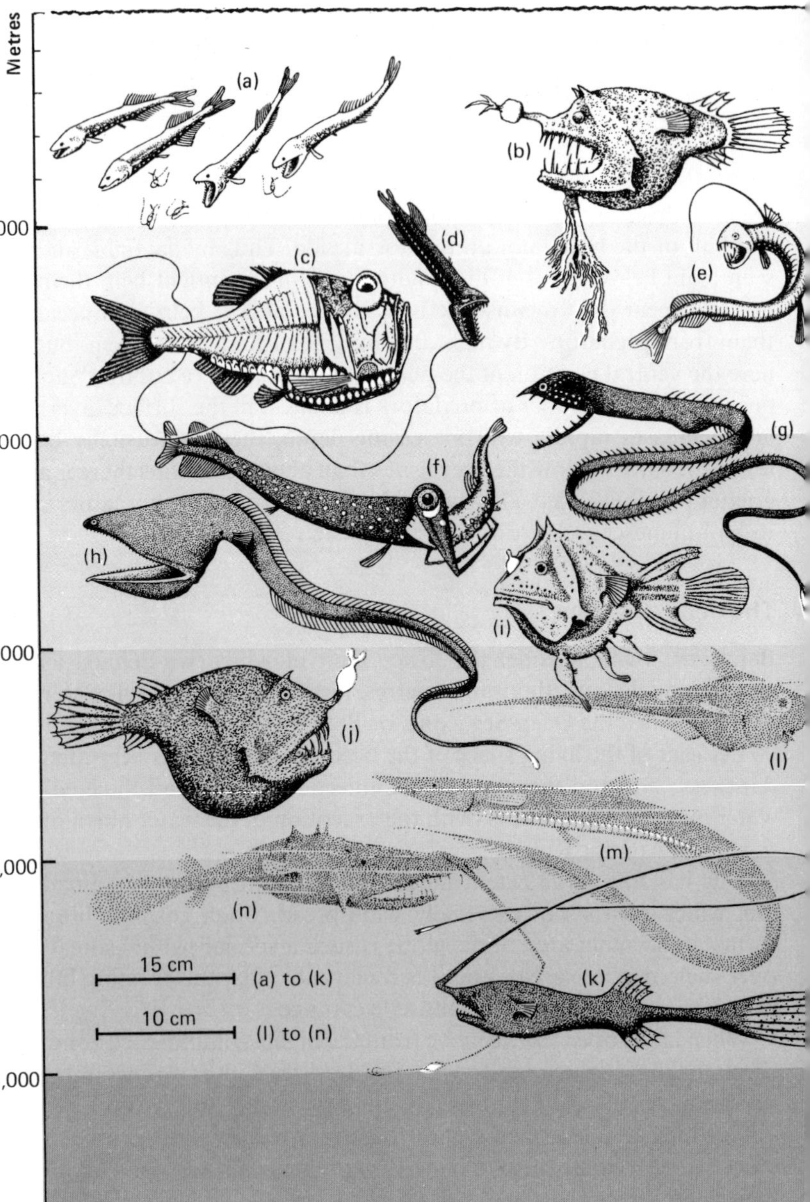

Metres
1,000
2,000
3,000
4,000
5,000
(a)
(b)
(c)
(d)
(e)
(f)
(g)
(h)
(i)
(j)
(k)
(l)
(m)
(n)
15 cm
(a) to (k)
10 cm
(l) to (n)

5·5 Fish inhabiting the deep sea zone. (**a**) Four *Cyclothone signata* hunting copepods; (**b**) *Linophryne arborifera*; (**c**) *Argyropelecus affinis,* with large telescopic eyes; (**d**) *Ultimostomias mirabilis,* with very long barbel; (**e**) *Chauliodus snoanei* (first part of dorsal fin thread-like); (**f**) *Malacosteus indicus* feeding on *Myctophum punctatum*; (**g**) *Saccopharynx johnsoni* (long teeth, whip-like tail); (**h**) *Eurypharynx pelecanoides* (large mouth, small teeth); (**i**) *Edriolychnus schmidti* (with three dwarf males attached to ventral side); (**j**) *Barophryne apogon* (with lantern on the nose); (**k**) *Gigantactis macronema* (long thread on nose); (**l**) *Macrurus globiceps* (with mouse-like head); (**m**) *Halosaurus rostratus* (with light organs on ventral side); and (**n**) *Lasiognathus saccostoma.* Most of the species are luminiferous. (Kai Olsen.)

Already, here, predatory species are found in much greater numbers and these reach their climax in the deep-sea depths. Although filtering copepods and shrimps living on detritus – or anything else – are still found here, a number of the shrimp species are purely predatory, and all the fish are specialised predators with highly developed equipment for catching their prey. These fish all eat each other and are constantly on the alert for enemies; they also compete with each other in impressive warning equipment against attack, and in terror apparatus designed to frighten an unknowing prey straight into a voracious mouth.

One of the commonest and most famous of these predatory fish is probably *Macropharynx longicaudatus,* which has an enormous throat resembling a funnel with a lid, while its body tapers to a whiplash tail – a living example of functional construction. When the stomach is empty *Macropharynx* appears very slender, but as it can expand to absorb anything that the powerful mouth can swallow, this fish (and the closely related *Saccopharynx*) is often found with a grotesquely distended body, digesting fish twice its size.

Another curious fish of this zone is the large toad fish, *Ceratias holboelli,* of which one of the most beautiful and largest specimens known is to be seen in the Zoological Museum in Copenhagen. The female can be over one metre long. Part of the dorsal fin is modified into what can be truly called a fishing rod, which slides forwards and backwards along the length of the fish, running in a canal in its back. In front of this fishing rod is a soft line with an illuminated bladder at the tip – the 'bait'. We have already seen how most sea animals are attracted by light (something used by Portuguese fishermen, among others, to great advantage – they rub the bait on their hooks in a secretion of luminiferous bacteria from the anal gland of the fish *Malacocephalus*) but millions of years before the method was utilised by man, it was developed and employed by deep-sea fish. We now know about eighty species of toad fish of thè *Ceratias* group, and apart from two species these all have a catching apparatus with illuminated bait. When the female of *Ceratias* lights her bait, other predators see this as a possible prey and approach. *Ceratias* then slowly hauls the fishing rod in along the canal on the back until the illuminated bait hangs temptingly, just above its large jaws. Just as the stranger is about to attack the bait, he is himself seized, to disappear into the wide mouth. The British biologist Colin Nicol, however, underlines the danger to predators of using such means: '... the showing of a light is a hazardous business in the deep sea, to be done at peril, and to be kept to a minimum.'

Ceratias is a remarkable animal in other ways. Its free-swimming, compressed larvae resemble those of many other fish, but already when only 4 centimetres long the larvae of males and females begin to differentiate. The female larva grows quickly, becomes more plump, and begins to form her 'fishing rod'. On the snout of the male larva a series of curved grasping teeth appear by which the larva attaches itself to the ventral side of a female. It thus abandons a free life and ultimately amalgamates itself with the female, so that during the rest of its life it is nourished by her. The male attains a length only one-twelfth of the female and had we not known it in its larval form we might easily have mistaken it for one of the anal fins of the female. Strategically placed near the sexual opening of the female it fertilises her eggs and this seems to be the only function it has.

Besides the abundance of fish this zone also seems to be especially rich in prawns and shrimps. Many of these species are reddish in colour, but in the deep sea where red light never penetrates they appear black and therefore invisible. Many deep-sea prawns and shrimps are provided with extra long, sharp rostral spines, no doubt weapons of defence; and other species, especially those of the genus *Sergestes,* have enormous antennae four or five times as long as their body. Deep-sea prawns of the genus *Acanthephyra* (which in their external appearance are very similar to the red deep-sea prawns fished and sold for food in Scandinavia and Greenland) can squirt out a cloud of luminescent fluid to confuse their pursuers in the same way as do some deep-sea cephalopods.

We have mentioned the 'inverted food pyramid' of which the free-swimming deep-sea animals are the apex, and we can see that the energy losses which occur at each stage of a food-chain result in intense competition in the deep waters. Not all the food potential, however, is used up by the bathypelagial fauna, as we shall see in the next chapter.

6 Life on the bottom

Life on the sea bed itself – the benthos – is much richer in species than is the pelagial. The pelagial population contains 3,000 species but the benthos consists of at least 157,000 species, corresponding to nearly 16 per cent of all living animal species on our globe. The great majority of bottom animals lives in depths of less than 200 metres, and there are very many more in tropical waters than in cold seas. In coastal areas the number decreases rapidly with depth; and in the deepest parts of the oceans (the hadal zone) we know of less than 370 species – less than two-thousandths of the total number of benthic species.

The Arctic and Antarctic coastal zones have much fewer species than the tropical zones, and it follows that temperature must be a very important factor. Thus it is reasonable to assume that the cold and warm sea currents must also play a considerable part in determining the number of species found in the different areas. If we look at the North Atlantic, we see that the rotation of the earth causes all sea currents to turn more and more to the right the further they are from their place of origin. The warm, north-moving Gulf Stream turns from the Caribbean towards Europe, where its warming influence is felt right up to north of Lofoten. In contrast the cold, south-moving Labrador Current presses in towards the Atlantic coast of Canada and the United States, so that coastal waters are cold as far as Cape Cod, which lies on the same latitude as the Bay of Biscay. Corresponding to this there are, for example, 465 species of prosobranchiate molluscs living along the west coasts of Europe, from the Murmansk coast to the Bay of Biscay, while the coast of America from Labrador to Cape Cod has only 143 species. The same conditions can be shown for other animal groups and in other seas with similar current patterns.

Epifauna and epiflora

It was mentioned earlier that the bottom organisms can be subdivided into epifauna and epiflora – which sit on the surface of the sea bed, or on rocks, stones, bulwarks, shells, eel grass, algae etc., and infauna – animals which live buried in the sand, mud and silt deposits which form the large plains of the sea bed.

The epifauna includes around four-fifths of all benthic animals. In polar seas the number of epifauna species is not much greater than that of the infauna, but the epifauna reaches its richest development in shallow tropical waters where the coral reefs in particular provide great opportunities for this kind of life. The most varied life conditions are found in the tidal zones of the warm waters, together with such a large choice of 'micro-landscapes' that almost every animal form can find a suitable place in which to settle. Also, it is only in shallow waters that we find large permanently attached algae in the form of sea weeds, which again, with their very distinct zoning, contribute to the creation of new micro-landscapes.

Finally, it is not known whether epifaunistic animals have had any chance at all to populate the rocky and stony areas which rear up in the very deep seas. We know that on the deep-sea bottom some rocky areas, for instance the intercontinental ridges, rise over the surrounding huge areas of silt like the 'nunataks' over the ice cap of Greenland. We also know that many of the deep-sea species lack a free-swimming larval stage and that their young crawl directly from the yolky eggs on to the bottom at the very place where the eggs were deposited by the mother. This means that such species have no means of dispersal. Epifaunal animals need a firm base on which to settle and therefore they have no means of traversing the huge silty deep-sea plains. How then would they be able to reach the isolated peaks? It must, however, be added that this is a purely theoretical question. We have, so far, no observations which would confirm or contradict it, because no research vessel has yet run the risk of lowering its expensive equipment on to the rocky peaks of the deep sea.

In the shallow waters along all the coasts of the globe we have not only an epifauna, but also an epiflora of seaweeds consisting of permanently attached green, brown and red algae, of which the brown are in a majority. These algae are of the greatest importance to animals of the shallow water areas. At ebb tide in the tidal zone the algae lie in a layer over the bottom, protecting the animals from drying out and from strong sunlight; many animals seek cover here and find favourable food conditions. They can also eat the alga itself or settle upon it, that is, use it as a permanent base.

Here we have an opportunity to show how an apparently simple combination of plants and animals actually consists of a very precise and delicate interaction. The large brown algae, for instance those of the genus *Laminaria,* often have their surface overgrown by crusts of moss animals (polyzoans) or by tufts of hydrozoan polyps, the so-called hydroids. These are, however, preyed upon by an abundance of sea slugs (nudibranchs), some of which feed exclusively on the polyzoans, while others specialise on a single kind of hydroid. The problem is that the polyzoans as well as the hydroids use the living leaves of the large algae as their base or substrate. The effective life of such an algal leaf is only about five to seven months, and during the two first months the leaf surface is usually covered by so thick a layer of mucus that most larvae, including those of polyzoa and hydroids, cannot settle on it. After this slimy phase of the leaf, the polyzoa and hydroids find a firm basis and start growing but it is not until after one or two months, when their colonies have grown up to a reasonable size, that they can keep step with the predation of the sea slugs. The larvae of the sea predators give up their free-swimming life to settle amidst their future food, and they find favourable conditions for rapid growth. This growth must of

6·1 Typical epifauna from the Australian tidal zone. Snails of the genera (**a**) *Littorina*, (**b**) *Patella*, and (**i**) *Thais*; the barnacles (**c**) *Cataphragmus*, and (**d**) *Tetraclita;* (**e**) a tube worm *Galeolaria;* (**f**) the mussel *Mytilus;* (**g**) a crab *Leptograpsus;* (**h**) a coat-of-mail shell, *Chiton;* (**j**) a sea urchin, *Heliocidaris;* and (**k**) a sea squirt, *Pyura*. (After E. C. Pope and P. M. McDonald.)

necessity be rapid, because the leaf will be cast off in about three more months; and during this period the slugs must attain full adult size, mate, spawn their egg masses on the algae and have their larvae hatched and developed. This is just one example of the hundreds of factors which affect the epifauna, but it demonstrates how precisely the seasons of growth and reproduction of different species must be balanced if the interaction between algal leaf, prey and predators is to function in the most economic way.

Eel grass Shallow water is not only the home of sea weeds, but also of flowering plants, in particular the eel-grass plants of the genus *Zostera,* different species of which inhabit protected shallow water areas along most of the world's coasts. Until the end of the 1920s the shallower areas of the Kattegat (Denmark) were covered by eel-grass meadows where, in late summer, the leaves were as dense as a field of corn and grew to several metres length, lying like long tapes over the surface of the water. The Danish marine biologist C.G.J. Petersen calculated that in the year 1910 there were 24 million tonnes of eel grass in Kattegat alone, corresponding to three times the entire Danish hay harvest for the same period. In 1931 an eel grass epidemic suddenly began. The disease started along the North American Atlantic coast and quickly spread to Europe and to Danish waters. The eel grass was afflicted with black stripes and spots, and both leaves and roots rapidly rotted away. The reason for the disease was never quite clear. Bacteria were believed at first to be responsible; but we are now more inclined to think that a primitive botanical fungus, *Labyrinthula*, was the actual cause. The curious thing is, however, that this fungus, which created havoc in the 1930s, still lives among eel grass, apparently without causing any damage. But, however it may have started, this disease of the 1930s was a catastrophe. It spread a few years later to the Pacific coast of the United States (it was thought that the fungus arrived there by way of oysters from the east coast of the United States which had been transplanted into California waters) and in a short while the eel grass along this coast too was badly attacked, in an epidemic which became little short of a major upheaval. Areas formerly covered by dense growths of eel grass were completely devastated and beaches which had been protected against waves were now mercilessly exposed to storms. Many small but important food animals, like the small gastropods of the genus *Rissoa,* formerly found in their hundreds of thousands in every cubic metre of grass, were deprived of their natural habitat and decreased greatly in number, and shallow-water fish lost both an important hiding place and a first-class hunting terrain.

In Danish seas the eel grass has never completely recovered. In fjords and sounds, where low salinity reduced the full impact of the epidemic, the grass was able to survive fairly well and has now fully regained its old terrain, but in the Kattegat, where salinity is higher, there are still large areas of bare sandy bottom where formerly eel grass grew in dense, waving meadows.

However devastating the epidemic was, it did help to change scientific opinion about conditions of production on the sea bed. It had long been known that the micro-algae of the free waters were the 'primary production' for pelagic animal forms, but there were still biologists who thought that the enormous amounts of dead and decomposing sea weed in shallow water areas must constitute by far the most important food source for bottom animals in coastal waters. Both Petersen and the Danish plant physiologist P. Boysen Jensen, who collaborated to elucidate production conditions on the sea bed, were very enthusiastic about this line of thought, which according to the knowledge of the time seemed both justifiable and reasonable. But with the disappearance of eel grass from large areas of our waters, biologists were presented with a completely unexpected demonstration on a really large scale. As mentioned, a number of small species directly connected with the eel-grass habitat suddenly became rather rare; but the animals living on the rest of the sea bed did not appear to decrease either in number of species or in number of individuals.

During the first four to five years after the catastrophe there were still a few biologists who did not feel convinced. They maintained that thick mattresses of slowly decaying eel grass were stored in the deeper parts of the Danish seas, and that this accumulation had not yet broken down into detritus and might still be taking part in the nutrient turnover. Now, however, it is nearly forty years since the large eel-grass meadows disappeared in the Kattegat, and there is still no indication that animal life on the bottom has been in any way affected. This, therefore, is proof that micro-algae, either living or dead, must constitute the most important food for all plant and detritus eaters living on the sea bed.

The eel grass is, however, not the only flowering plant of quantitative importance. In the Mediterranean we find extensive meadows of *Posidonia caulini* with leaves six to eight centimetres broad and up to one metre long. In the tropical Atlantic we find luxuriant beds of *Thalassia testudinum,* the so-called turtle grass, which has somewhat broader and shorter leaves. Both these grasses are teeming with life. In the *Posidonia* meadows well over one hundred different species of animals have been counted, while a well developed *Thalassia* bed may contain 30,000 individual animals per square metre. It is also known that 113 different species of algae may use the *Thalassia* leaves as a habitat. Such algae are called epiphytic, which means that they live on the grass but do not draw any nourishment from it.

The infauna

There are far fewer species of animals living *in* the sea bed – only about 30,000, as compared with an epifauna of over 125,000 species. It appears that the number of species in these smooth sediment bottoms is not much reduced in polar areas as compared with tropical waters. But even though most sediment bottoms are comparatively poor in species they can well be rich in individuals, especially on banks in coastal waters. Unlike the epifauna, some of which lives on rocks, it is much easier to undertake a quantitative reckoning of the number of animals on these relatively smooth bottom surfaces.

A steel-jawed bottom sampler invented by Petersen is able to bite up one tenth of a square metre from the sea bed – and bring the sample to the surface for examination. The bottom material is then sieved and the animals washed, sorted, counted and weighed. When

6·2 Parallelism in the weight per square metre of benthic infauna in a high-arctic and a tropical fjord. (After G. Thorson 1957.)

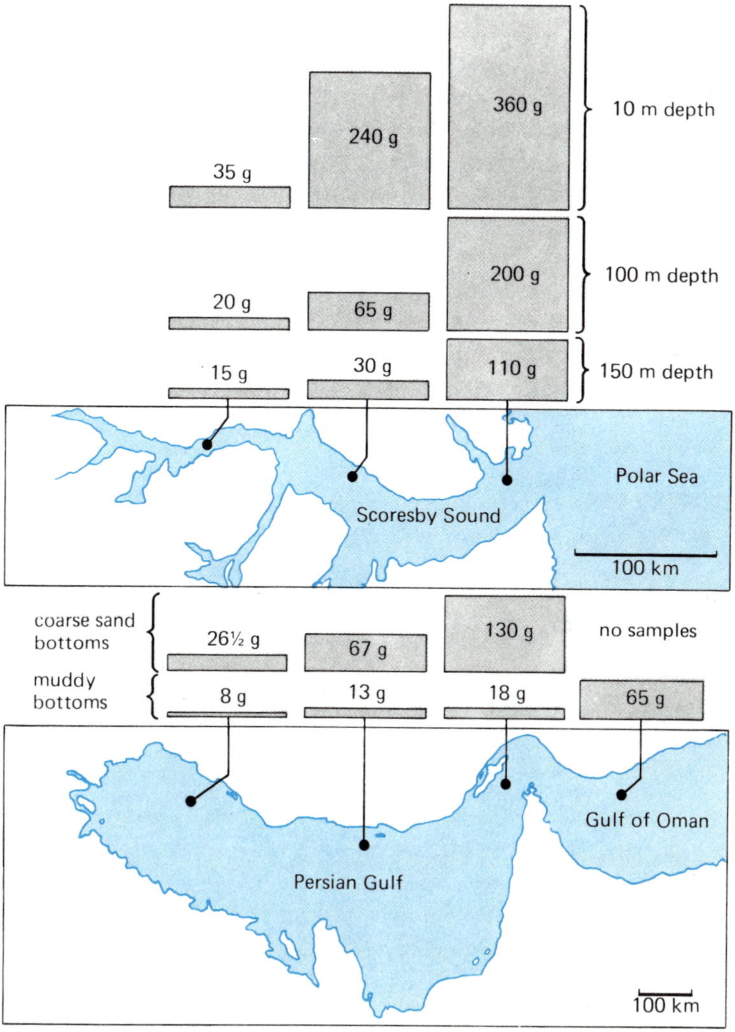

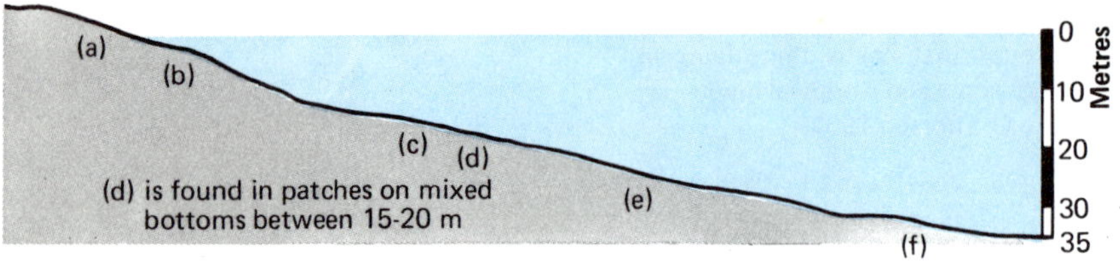

(a) Shore fauna

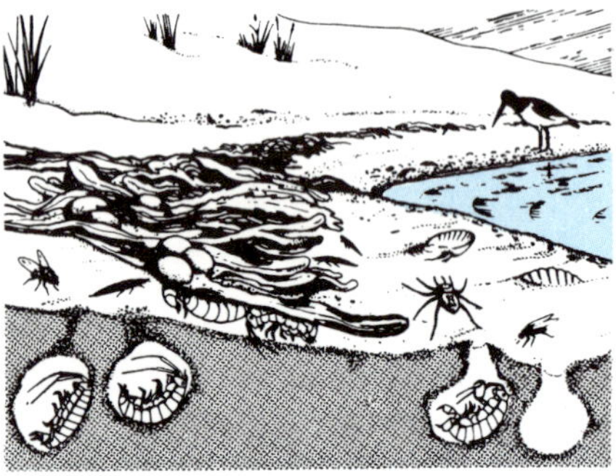

(b) Macoma community

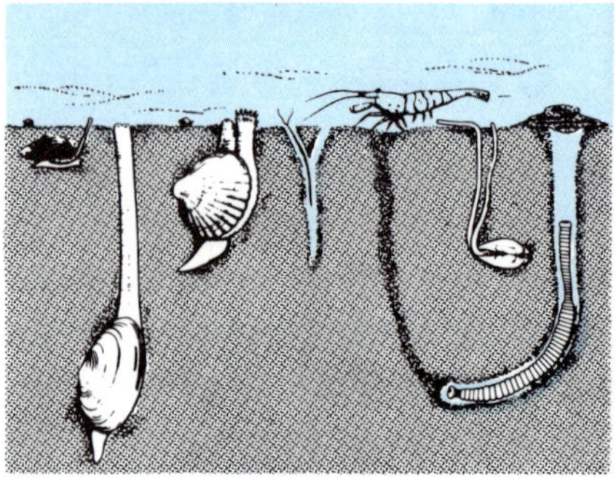

(c) Venus community

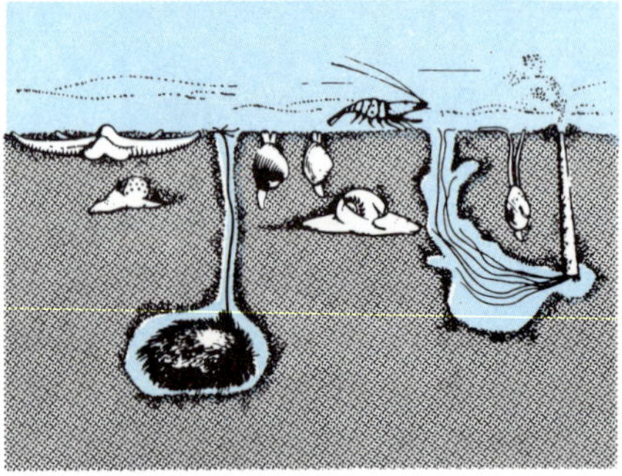

(d) Syndosmya community

(e) Amphiura community

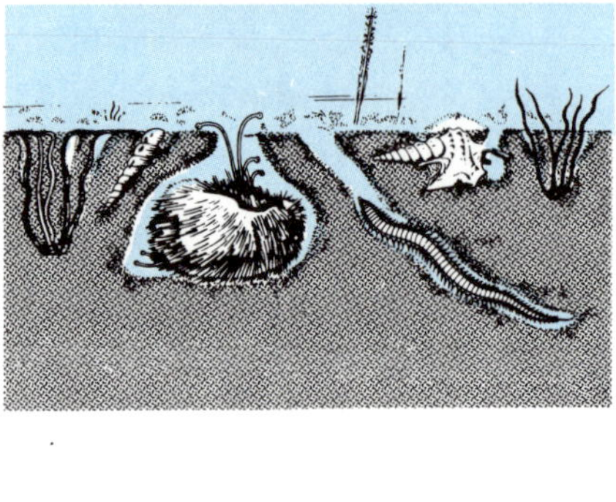

(f) Haploops community

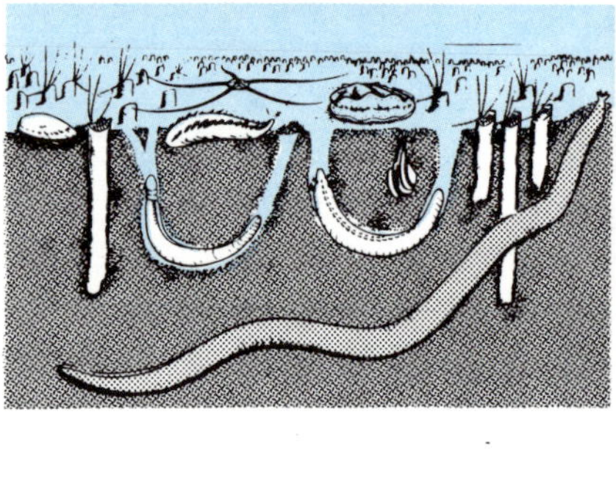

6·3 A section through the infauna communities at different depths in the Sound between Denmark and Sweden. (After G. Thorson 1968.)

working in temperate seas, it soon becomes evident with this method that in every sample there are only a few really plentiful animal species, and that these few species are dominant over large areas with the same type of bottom. If the bottom sampler is used on another type of bottom, different animal species are often found in the sample, but again only a few species occur in large numbers; with repeated samples these can be regarded as *type animals* for that particular bottom. After having taken thousands of quantitative bottom samples in Danish waters, Petersen felt convinced of the regularity of the observations, and between 1911 and 1918 established his well-known theory of animal communities, which led to a new era in marine bottom investigations.

We now have tens of thousands of quantitative bottom samples from almost all over the world. Many nations have adopted Petersen's method and have collected a large mass of material. We also have samples from depths of one metre to over 10,000 metres (the latter taken by the *Galathea* Expedition in the Philippine Trench and later by Russian expeditions). Although Petersen himself attempted to establish a provisional map for the animal communities in the North Atlantic on a purely theoretical basis, he would certainly be surprised to see how much the information we now have can tell us. In arctic, cold-temperate and some warm-temperate waters, it has been shown that practically everywhere in coastal areas one can find parallels to the smooth bottom communities which Petersen described from the area between the Baltic and Skagerrak. On the same type of bottom at about the same depth, irrespective of the latitude, we have parallel animal communities, where different species, but of the same genus, are dominant in the samples.

This distribution of dominant animals by depth is very precise. For example in the tidal zone, where the bottom is uncovered at low water and the sea bed consists of mixed mud and sand, we find everywhere a dominance of mussels of the genus *Macoma*. On a pure sand bottom at 10–20 metres, mussels of the genus *Venus* are dominant. On mud bottoms from 15–20 metres depth there are large numbers of the brittle star *Amphiura*. We can therefore refer to the *Macoma, Venus* and *Amphiura* communities of many different seas, and where such parallel communities occur we are in a position to single out which of the bottom animals from our present seas are so dominant that they will in the future become guide-fossils of the geological layers from our period – as were belemnites, trilobites, ammonites and brachiopods for earlier ages in the history of the earth. The concept of parallel communities seems, however, to apply only to cold-water fauna, breaking down when we consider bottom fauna in tropical and some warm-temperate regions. In these regions we find a large number of *species* but only a few *individuals* of each per square metre. There is so far no satisfactory explanation of this.

Territories

It is interesting to consider the area of territory that individual benthic animals need in order to live. When animals from a bottom grab sample lie freshly washed in a sieve they often look rather unimpressive, and if we visualise them evenly spread in their one-tenth of a square metre we get the impression that the population is surprisingly small. But we must remember that animals in a sieve are in a state of shock, owing to the rough treatment to which they have been exposed. Mussels and snails, for example, withdraw into their shells, closing themselves off from the outside world as much as possible; bristle worms contract and hide in their tubes. But if we take a bottom sample and, without sieving it, carefully place it in an aquarium, we obtain a very different impression of life on the sea bed. The detritus-eating mussel species, for example, open their shells and stretch two proboscis-like siphons from their posterior end. One siphon is short and serves as the outlet for the respiratory water. The other is at least twice as long as the mussel itself and, like a flexible vacuum cleaner hose, it swallows detritus from the bottom all over the radius of its reach. The 6 to 7 centimetres long north-west American bivalve *Macoma secta* can extend its inhalant siphon to more than 1 metre length, fifteen times the length of its shell, and may sit 45 centimetres deep in the bottom. The unimpressive tube-dwelling bristle worm *Terebellides,* one of the type animals of the mud bottom in most of the earth's coastal waters, is, when contracted, only 4–5 centimetres long. Seen from above with the front of the body just level with the bottom it will, at most, cover a surface of 0·5 square centimetres, but fifty fully expanded individuals, eight centimetres long, with tentacles spread over the bottom, are able to collect food from an area almost four hundred times larger. Consequently, when there are fifty *Terebellides* per square metre, as often happens, their tentacles cover the whole of the bottom surface and monopolise the entire area as a food source.

The territorial demands of unicellular animals are even more surprising. In the deeper waters of the Arctic seas, and in certain areas of the North Sea and Kattegat, one finds around fifty foraminifers of the genus *Astrorhiza* per square metre. For unicellular animals these are large – about 5 millimetres in diameter – but their shells, which are composed of sand grains cemented together, take up very little room in a bottom sample. In life, however, long slimy threads of protoplasm emerge from a multitude of holes in between the sand grains, threads which wreathe around each other to form an irregular network over the bottom. This net of protoplasm has been compared to a spider's web, but with the difference that the protoplasm of foraminifers is both spider and web at the same time. The threads are sticky and can catch and digest amphipods of up to three centimetres in length. *Astrorhiza's* net can cover an area with a radius of 6–7 centimetres. In other words, even if only fifty of these inconspicuous creatures are found per square metre, over half the bottom surface will be a danger zone for the passage of many small crustaceans and young bottom animals.

PART TWO

7 The supralittoral and littoral zones

The supralittoral zone

In the following chapters we shall discuss the plant and animal life of the sea bed, zone by zone, beginning with the beach and tidal zones and ending with the deep sea bed at over 10,000 metres. The supralittoral zone is the area just above high water mark, covered only by the sea in time of storm or flood. What we normally call the beach is also part of this zone and our reason for including it is that the spray from the waves may in some localities leave its mark on the life in this area; also the heaps of sea weed cast ashore make a basis for a considerable part of the animal life there.

On steep rock surfaces against which a heavy swell breaks, as for instance in the Faroe Islands, the sea spray can reach up to thirty metres above high water mark, leaving enough moisture for periwinkles to live there. On rocky coasts a very high tide or a storm wave can fill crevices high above the sea surface with salt water. From a biological point of view these rock pools are of great interest, as the organisms that inhabit them can be subjected to vastly differing conditions. The sea water which fills them up during a flood wave may gradually evaporate, thus attaining a salinity much higher than that of the open sea, or it may be diluted with rain water, so that it becomes brackish. Observations on what type of animals can survive and flourish in each single rock pool can, therefore, teach us a great deal about the tolerance of individual organisms.

On stones and rocks in this zone we also have a series of characteristic species of animals, which thrive nowhere else. Along most of the European coasts primitive insects of the genus *Machilis* may be seen darting away over the black crusts of lichens covering stones or rocks; but in warm-temperate and still more in tropical regions the most conspicuous animals on these stones will often be 3–4 centimetre-long isopods belonging to the genus *Ligia*. Along the coasts of, for instance, Guinea in West Africa, they are extremely abundant, and in houses built on the shore the inhabitants have the doubtful pleasure of occasionally seeing one of these large crustaceans crawling up and down the walls of their living rooms.

In the supralittoral zone insects play an important part. On the sandy beaches and in the dunes it is usually small beetles of the genus *Bledius* which predominate. Among the decaying masses of sea weed cast ashore, a rich and varied animal life is on display. One Swedish thesis lists more than five hundred species of small animals from this zone, among them being ten species of amphipods, thirteen of millipedes, forty-four of spiders and more than three hundred species of beetles. Most of these animals are invisible until the heaps of sea weed are disturbed; the species with which one may be familiar during a normal excursion along the beach will normally be confined to some of the more rapid spiders, *Trochosa* and *Tarentula,* the audacious beach flies, *Fucellia,* the larvae of which develop in the rotting sea weeds, and the amphipods *Orchestia* and *Talitrus*, which especially during the twilight hours may be seen in great numbers jumping on the beach.

The beaches and shores of subtropical and tropical regions have another important faunal element, which those who have seen will hardly forget. These are the large herds of crabs. In the Persian Gulf I have had the opportunity to observe herds of the crab *Ocypode* (the ghost crab or racing crab) at noon and in full sunshine. At first, when I saw a herd of about a hundred of these animals at a distance, I took them for a flock of sand pipers. Their colour was yellowish-brown, and they all moved like lightning in the same direction, with military precision and with responses so rapid that they could hardly be followed by the human eye. I watched them through my binoculars, and cautiously approached the sand bank on which they lived. As if by magic they all disappeared; in less than one second they had dug themselves into the sand, with only their large eyes visible at the surface. Again I approached slowly, still using my binoculars.

7·1 Moon-phase spawning grunion (*Leurestes tenuis*) on a Californian beach. (After B. A. Woodling 1968.)

Suddenly the eyes disappeared and only the very tips of the antennae were visible. When I finally reached the site the antennae too were withdrawn. I then tried to dig out the crabs one by one, but failed; it was also impossible to chase them as they ran faster than I could. It is no wonder that such crabs can fully dominate all other animal life on a tropical sand bank if they occur in large enough herds.

The littoral zone

The area between high water and low water is called the littoral or tidal zone. The extent of this zone is governed by more than one factor. The first is the fortnightly variation in tidal range, from a high and low maximum (spring tide) shortly after full and new moon, to a minimum range (neap tide) a week after each spring tide. The second is the local configuration of coast lines, which has a profound effect on the tidal range. For example, on the open coastline of the Gulf of Maine, on the Atlantic coast of the United States, the range is no more than a few decimetres, while in the inner part of the Bay of Fundy (a bottleneck) it is as much as fifteen metres. Furthermore, the frequency of the tides varies. In the Atlantic the tide rises and falls twice in every 24 hours and 50 minutes, but there is only one tide per day in the Gulf of Mexico, while the Mediterranean and the Indian Ocean exhibit a mixture of twelve and twenty-four hour rhythms. Landlocked seas such as the Mediterranean and the Baltic have a tidal range of about 30–50 centimetres. It will be seen, then, that the tidal zone varies greatly from place to place, while the rhythm and intensity of tidal action must have a profound effect on the fauna and flora.

Moon-phase spawnings While on the subject of tidal rhythms, it is appropriate to mention the connection that exists between breeding, tides and even the phases of the moon. These phenomena are

referred to as moon-phase spawnings (lunar periodicity) and here are two of the most celebrated examples.

The first time I visited my marine biologist colleagues at the Scripps Institution of Oceanography in southern California, I was, much to my surprise, invited to catch fish on the shore at around 10.30 p.m. I naturally thought this was a joke, and my suspicion was strengthened on learning that we were to take only a bucket and a pocket torch. This was in March, at full moon and high spring tide, and we were soon on the wide sandy beach watching the long rollers curl and break on the shore. The beach was milling with people, and suddenly we heard the cry, 'Here they come'. Ten minutes later the scene became one of great activity. Out of the breakers leapt large numbers of silvery fish, approximately 15 centimetres long. They usually came out of the waves in pairs, a male and a female, and swam or slithered with amazing speed and purpose over the wet

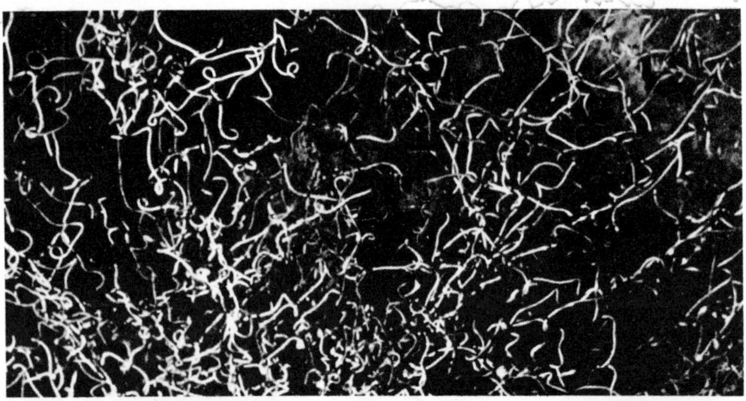

7·2 Moon-phase spawning in Samoa of the Palolo worm *Eunice viridis*.
Top swarming Palolo worms at the surface of the sea at 0200 hours.
Centre High Chief Le'iato catching Palolos from his canoe.
Bottom Freshly caught Palolos filling a palm-leaf basket.
(After C. Hauenschild *et al* 1968.)

sand towards high water mark. If one was quick enough to shine the torch on them during the next few seconds the female could be seen burying herself in the sand, covered by one or more males. Here the wedding took place, the eggs were laid and fertilised, and in about thirty seconds the spawning fish returned to the sea, borne by the next large wave which, at the same time, brought new bridal couples on to the beach. On spawning nights these little fish, called grunion (*Leurestes tenuis*), lie so closely packed together that the whole beach is a mass of silver, and it is impossible to move without treading on them. If one is quick enough, the bucket can soon be filled with fish, but even though the beach is swarming with people collecting as many as they can, a great number reach the safety of the sea again with the next wave.

The eggs are laid, fertilised, and covered with about 5 centimetres of damp sand. As they are laid at the high water line of the spring tide they will not be exposed to the waves but can develop in peace in the warm, sun-baked sand. Nine days after spawning the eggs are ready to be hatched, but hatching does not take place until next spring tide, five days later, when the waves again reach far enough up the beach to disperse the sand and carry the eggs out to sea. Once the sand is removed, the young hatch in less than three minutes. But if for some reason the water does not reach them two weeks after spawning, the eggs remain in their capsules for another two weeks, or even four, until a wave finally reaches and liberates them. In spring these grunion spawnings can be seen every fourteenth day, always at high water, always at night, and with such precision that one can predict ahead exactly when the spawning fish will appear.

Off the Islands of Samoa and Fiji we meet another amazing example of rhythmic spawning, known and exploited by the natives for hundreds of years before coming to the notice of zoologists. At daybreak, in the months of October and November, and on the day before the moon is in its last quarter, the inhabitants of these islands set out for the beach with baskets, buckets and dishes. With incredible precision, just before sunrise, the first macaroni-like worms appear on the surface; a few minutes later the water will in places resemble noodle soup, so numerous are the worms. Actually, these are the hind parts of the palolo worm, *Eunice viridis*, a polychaete worm which spends its life hidden among corals in shallow water. As the time of sexual maturity approaches, the hind part of the animal changes shape; it becomes thin-skinned, slender, with special swimming bristles. The hind part of the female is filled to breaking point with green eggs and that of the male with yellowish-white sperm. When the time is ripe, the hind parts are released from the rest of the body, and while the forepart unconcernedly continues its life among the coral, the hind parts swarm towards the surface Here they burst open and eggs and sperm are scattered in the water. It is just before they burst that they are collected, for they are considered a great delicacy, and as they swarm so densely in the water they can almost be scooped up. This whole spectacle takes place in only a few hours, but with such precision that the inhabitants of the islands know exactly which morning the swarms will arrive.

However, it is not necessary to go to California or Samoa to find moon-phase spawnings. Even in Danish seas, for instance in the Kattegat, where the tide is rather weak, a rhythmical spawning takes place every spring which is nothing short of astonishing. On the low sand flats, from high water mark down to four metres, lives the bristle worm (a polychaete) *Nereis virens*. Normally, very little is seen of this animal as it is buried in the sand, but if one hauls a small piece of fish slowly along the bottom, the front end of the worm will often push up to seize the prey with its large jaws. Not until spawning times does one discover how numerous it is and how impressive its size. *Nereis virens* can reach a length of 70 centimetres, equal to the length of a medium-sized adder; it is without doubt the largest polychaete found in Scandinavian waters. During the breeding season its bristle feet turn into webbed feet; the body is filled with eggs or sperm, and at full moon in April these large worms swarm in the water, spawn and then die. The fishermen in Kattegat call them leeches, and are very familiar with this recurrent yearly phenomenon. During particularly large moon spawnings the swarming nereids can fill the net, entwining themselves in the mesh in such numbers that they are a nuisance to fishermen. While swarming is in progress one can see large numbers of *Nereis* being eaten by sea gulls, and larger sized plaice stuff themselves with so many of these spawning bristle worms that the stomach can be seen as a visible bulge on their flat bodies.

Similar moon-phase spawnings are known for a great number of sea animals – fish, many different bristle worms, oysters, common mussels, periwinkles, sea urchins and medusae. Since the phases of the moon in many places will coincide with flood- and ebb-tide, it is difficult to decide whether the spawning is started by the direct attraction of the moon, by the light which it reflects, or by the height of the water. The example from the Kattegat, where there are hardly any tides at all, seems to point to an influence of the moon itself. Curiously enough, it is not the same phase of the moon which influences the swarming of all species. Some species spawn at full moon, other at new moon, others when it is in its first or last quarter.

Two things about these rhythmical spawnings have particularly intrigued biologists. The first is: what do the animals achieve by this phenomenon? The advantage of all individuals spawning in the same hour is obvious, because a species which sheds its sperm and eggs freely in the water has a much greater chance of 100 per cent fertilisation when large amounts of eggs and sperm are spawned simultaneously. The second question is: what is responsible for setting the rhythmical spawning in motion? Is it the moonlight itself; or tidal ebb and flow; or is it a third factor wholly or partly influenced by one of these?

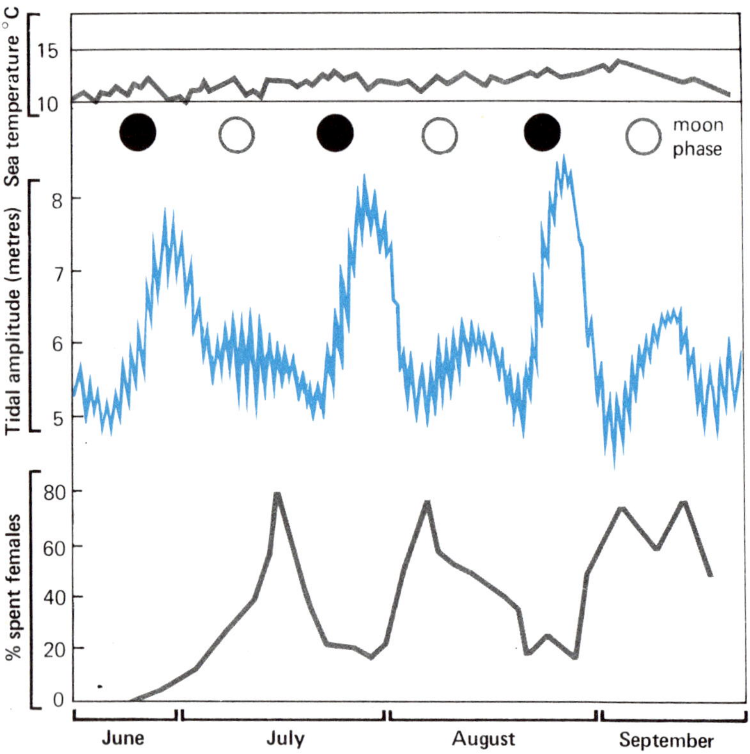

If marine animals that in nature show a marked lunar periodicity are placed in a dark room in large, well aerated aquaria of the right salinity and temperature, they often keep their rhythm for a long time, even though they are isolated from both moonlight and tidal rhythms. The reason for this may, however, be that they have the lunar or tidal rhythm 'built in' so to speak. In order to get decisive proof of in-built rhythm, it would be necessary to carry the experiment on through several generations of captive animals and see whether the rhythm persists or dies out. Such an experiment would be very elaborate and costly, which is perhaps why this problem has not yet been solved.

7·3 The moon-phase spawning of the common mussel *Mytilus edulis* on a North Atlantic coast. (From H. Battle, after G. Thorson 1950.)

Diurnal rhythms The activities of many animals are clearly connected to the daily tidal movements; we come across an amusing example on the flat sands submerged at high tide in temperate and tropical coastal areas.

On hard sand beaches which receive the full impact of the waves from the open ocean we often find small bivalves of the genus *Donax* in immense numbers. For example, in one belt of the tidal zone in southern California, 2–5 metres wide and 10 kilometres long, about 20,000 individuals of *Donax* were found per square metre. These wide, hard sandy beaches provide what would appear to be an extremely difficult habitat for a bivalve. At low tide the sand is often so compact that it is hardly possible to make a visible footprint; but each time the heavy breakers roll in the sand is shifted, disturbing the animals living there. *Donax*, however, seems actually to depend on strong water movements, created when the waves break on the sand; and as the tide rises and falls *Donax* moves in rhythm with it. At low tide, and just before the flood tide starts, all the *Donax* individuals are near low water mark and completely buried in the sand. When a wave of a certain strength arrives, one sees hundreds of *Donax* individuals, which literally jump out of the sand, being carried further up the beach. Then, when the wave is nearly spent, the animals bury themselves in the sand. While in transit, *Donax* trails its foot like a kind of floating anchor, and is able to dig itself in completely in a few seconds. With each successive wave it repeats this process until it reaches high water mark. But when the tide ebbs *Donax* promptly changes its tactics. It now lets each wave pass over, and jumps only when the water is in retreat. The process is now reversed and *Donax* gradually travels back to low water mark with the falling tide. These journeys up and down wide sand beaches can often cover a total distance of thirty metres each way, quite an impressive achievement for a 3 centimetre bivalve. We know that, on a rising tide, each jump is triggered by an acoustic stimulus – by

7·4 Vertical section of a West Indian beach, showing in six phases how, within a few seconds, the clam *Donax denticulatus* burrows into the sand. (Redrawn from B. A. Wade 1967.)

the vibrations of a breaking wave crashing on the beach. We know, too, that *Donax* responds only to the breakers of a certain weight and that one can trick it into popping up prematurely by stamping hard on the sand. But it is not fully understood how *Donax* responds in quite an opposite way when the tide is falling, delaying its jump until a wave is pouring back. Perhaps it may be due to an internal 'time-sense' in the animal itself, but it may also be caused by external conditions. When the flood wave sets in, *Donax* is buried in dry, relatively compact sand; when the ebb tide begins it is hidden in loose sand soaked with water. It may be that these conditions influence the acoustics and stimulate *Donax* to two different reponses.

Another common inhabitant of the tropical sand flats, the turreted gastropod *Terebra*, exploits the tides in a different way. On a rising tide it simply crawls over the sand towards the high water mark and digs itself in, but when the tide starts to fall and a large wave is returning, *Terebra* pops up out of the sand in one jump, spreads its large foot like a sail, to be seized by the withdrawing water and passively carried down towards low water mark.

Tidal exposure Many animals in the tidal zone are sessile, that is, they are permanently attached to a surface, which means that at low tide in a tropical climate they must endure the burning rays of the sun. For example, there are the acorn barnacles *Balanus*. These small crustaceans, like tiny white volcanoes, cover rocks and breakwaters in the tidal zone with cone-shaped calcareous rings. When these are exposed to the air, they change their method of breathing. Under water they breathe with gills, but they are able to 'switch over' to air breathing when necessary, which takes place via a very small opening and with the smallest possible loss of water. Although one of the Scandinavian species, *Balanus balanoides*, is a sea animal with pelagic larvae, we know that it can live out of water for as long

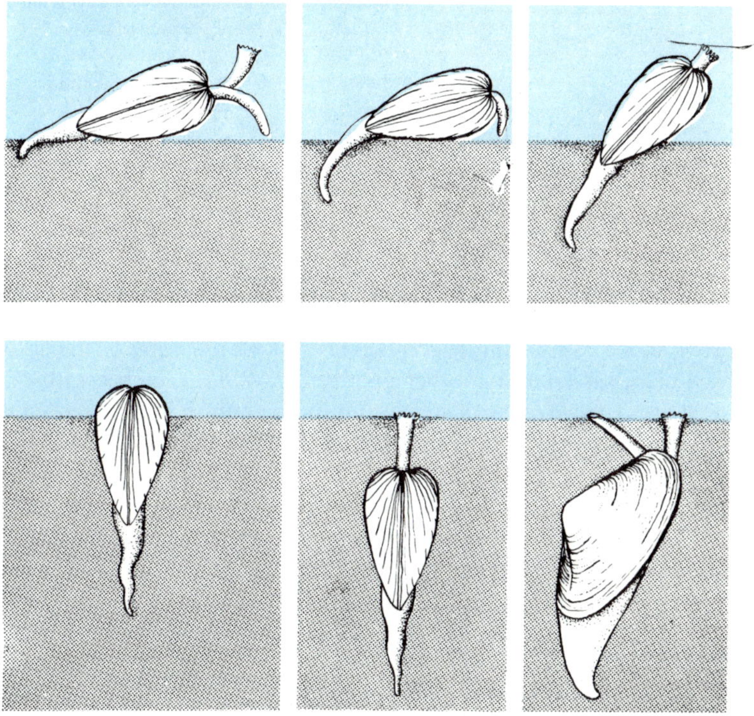

as six weeks; it immediately opens and begins to take food as soon as it is again submerged by water. Another, still more robust European species of acorn barnacle, *Chthamalus stellatus,* survived three full years on a laboratory table; it was submerged in sea water only one or two days a month.

Some tropical gastropods living between tide marks have developed an almost incredible resistance to desiccation. They can close their shell aperture with a horny or calcareous cover, the operculum, and even if the temperature of the sun-heated stones on which they sit reaches high levels they can survive out of water for one and a half years or more. Many other animals of the tidal zones can avoid desiccation only by digging down or seeking shelter in shady, humid places.

But heat and desiccation are not the only hazards for intertidal

7·5 The wastage during development of the barnacle *Balanus balanoides*. Only 0·2 per cent of all larvae hatched from one female are alive when metamorphosis sets in. (After G. Thorson 1968.)

animals. In the polar regions and even in cold-temperate zones they must be able to survive a frozen winter. Observations made on Scandinavian coasts tell us that several species that can move quickly simply transfer themselves into deeper water when the temperature approaches zero. Others, less mobile, must stick it out and these prove to be astonishingly robust. The cockle (*Cardium*), the soft-shell clam (*Mya*) and another bivalve (*Macoma*) together with the lug worm (*Arenicola*) can survive up to one month in a frozen sand bottom. *Mya* can even be frozen to −4°C for seven weeks without being killed. Nevertheless, unusually long and hard winters can be fatal to many inhabitants of the tidal zone. Often, it is not the low temperature itself which hurts the animals. A sheet of thin ice may cover the surface of the sand at low tide, thus preventing them from receiving fresh oxygen and leading to suffocation.

Epifauna of the littoral zone

Periwinkles and barnacles At the edge of high water mark or even higher the periwinkles are the dominant animals. They can be very close together, but even when they are sitting on freely exposed rock surfaces they use every crack to take shelter facing the light and the surf. The largest genus is *Littorina,* of which there are representatives all over the world, from the ice-covered rocky coasts of Greenland to the sun-baked reefs of the tropics; but many other genera are known, especially in warm seas. Common to them all, and requisite for existence in the tidal zone, is a thick, solid shell able to withstand the roll and force of the waves. The common North Sea periwinkle, *Littorina littorea,* occurs as frequently as 230 to 950 specimens per square metre, while a smaller, viviparous species occurs at Plymouth up to 10,000 specimens per square metre.

Below the periwinkle layer there is often a sharply demarcated belt of acorn barnacles – small crustaceans which after a two-month

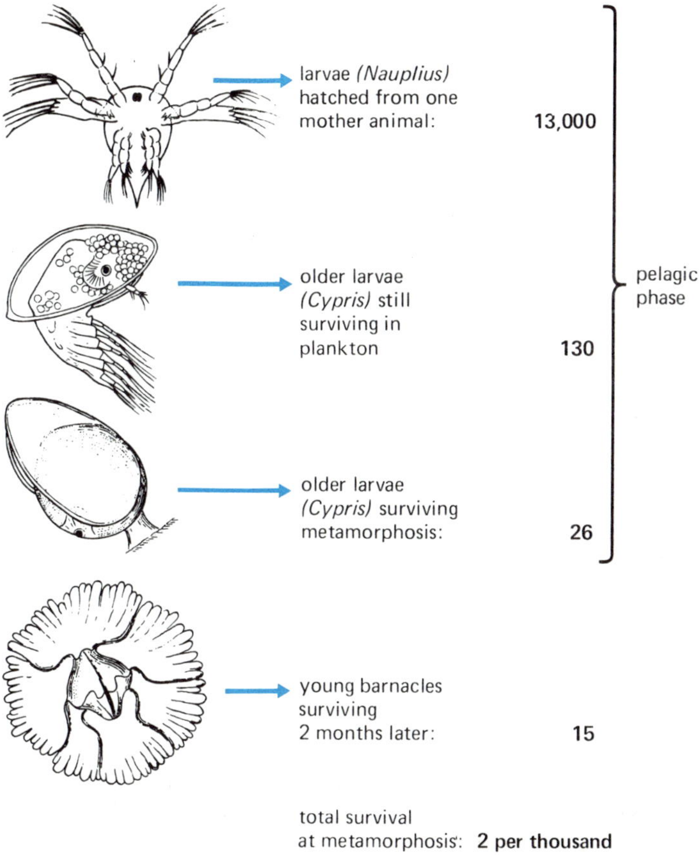

free-swimming larval life become permanently attached to rocks or stones, build a cone-shaped calcareous ring round themselves, and are walled in for life. They lie on their backs and comb the water with their cirri (specially adapted legs) which protrude from the mouth of the crater in their volcano-like armour. Even on coasts with very strong tidal waters the barnacle belt is hardly more than two metres wide, and often narrower. Inside the belt, however, the barnacles may sit in thousands per square metre. There can be several different species, each forming its own horizontal belt on the rocks. As the belt is narrow, as the barnacles must find their way there at the end of their pelagic larval stage, and as the choice made when they finally attach themselves is 'for life', it is understandable that larvae seek their permanent habitat with the greatest precision and thoroughness. In the case of a common European barnacle, *Balanus balanoides,* it has been estimated that of approximately 13,000 larvae produced by a single female, barely thirty individuals survive to attach themselves to a suitable habitat, and of these only fifteen will be alive two months later; the species takes two years to become sexually mature, so it is important that very few individuals die after settling permanently if there are to be enough left to maintain the population.

We know that the larvae of tidal-zone barnacles are photopositive, that is, they move towards the light and the surface of the sea, until the time comes when they settle on the rocks. It is only by staying right at the surface that they get a chance to encounter rocks and stones in the tidal zone. When they finally come in contact with a rock or a stone, they first crawl around on its surface, examining it in detail. At this stage the crawling barnacle larvae suddenly turn over to being photonegative; they are attracted to shady places, preferring dark coloured stones to light ones, and cracks or crevices to smooth surfaces; and they finally settle at a fairly precise angle to the predominating water current.

Even when the barnacle larvae are lucky enough to have all these demands satisfied, there still remains one important factor if the population is to survive. Barnacles are hermaphrodite – both male and female in the same animal – but self-fertilisation does not usually take place. When the adult is inexorably fixed to its chosen spot, it must have other individuals in close proximity if fertilisation is to occur. This is all the more important as barnacles do not spawn eggs and sperm freely into the water; on the contrary, the eggs are fertilised in the mantle cavity of the female by mating with a close neighbour. The sexual organ can only reach twice as far as the diameter of the crater surrounding the animal, so it is essential that the animals be close together. Barnacle larvae experience a strong attraction, supposedly of a chemical nature, towards permanent individuals of their own species. If two slates are suspended in sea water, one smooth and free of growths and the other covered with barnacles, the larvae attach themselves to the second slate, ignoring the first. Even if we scrape off the adult barnacles so that only their basal plates remain, most of the larvae are still attracted to the places where the adults had been located. This phenomenon is called gregariousness. A similar tendency is found in a number of other sessile animals which are hermaphrodite. As long as one permanently attached animal can attract another of the same species, they can cross-fertilise and ensure the next generation.

Although the advantage of gregariousness is obvious, there is also a drawback. So many young specimens may be induced to settle among the adults that their density can be fatal when they grow up. Densities of 600,000 to 700,000 young specimens per square metre have been observed in *Balanus improvisus,* but a fully grown individual needs a living space of about 80 square millimetres, from which it can be calculated that only 12,500 adults can live inside 1 square metre. This means that only 1·8 to 2 per cent of the young that originally settled can grow to adult size; the remainder have no room to develop, and perish.

Predatory snails Below the barnacle belt, in the lowest part of the tidal zone, life on rocks takes on a much richer character. Here animals are covered by water nearly all the time, and the purely marine milieu is beginning. Here we meet the predatory snails which have already begun to hunt the lower part of the barnacle zone. These snails have a supple proboscis which is inserted through the barnacle's crater; at the tip of the proboscis there is a rasp with a band of horny backward-pointing teeth which can quickly tear

7·6 **Below** The predatory snail *Purpura lapillus* on a rock in Yorkshire, England. It can be seen that nearly all the barnacles, the favourite food of *Purpura,* have been eaten by the snails. (After J. R. Lewis 1964.)
7·7 **Opposite** The common whelk *Buccinum undatum* feeds on living prey as well as on carrion, which it scents as it moves its long siphon in all directions. (After G. Thorson 1968.)

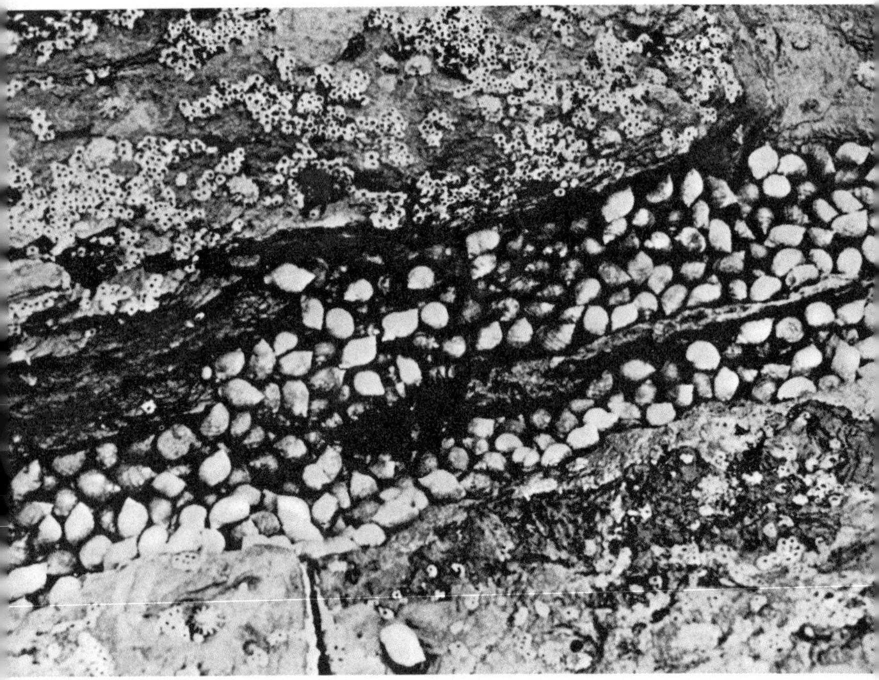

into the soft interior of a barnacle and transport the food to its mouth. In most seas, snails of the genera *Purpura* and *Thais* are dominant in this part of the tidal zone. As well as attacking barnacles, they are one of the most dangerous enemies of the common mussel. The snail's rasp-like teeth can bore a cylindrical hole through the shell of a living mussel and scrape the soft parts out.

The European species *Purpura* (*Nucella*) *lapillus* specialises in either barnacles or common mussels and concentrates solely on one of these. In one tidal zone in France, all *Purpura* snails were found to be feeding on barnacles. The barnacle population was gradually

reduced while that of the common mussel in the same zone correspondingly increased. As long as the slightest trace of barnacles remained the mussels were left in peace, and it was a long time before the snails learned the technique of exploiting mussels instead. Some years later the picture changed. The barnacles increased while the mussel population decreased, but the predatory snail had now accustomed itself to a diet of the latter and returned apathetically and unwillingly to a barnacle diet. Furthermore, we know that the snails' diet has a direct effect on the colour of their shells. Specimens of *Purpura* feeding on barnacles have yellow-white shells, while those feeding on mussels develop shells the colour of egg-yolk. Thus, simply by observing the shell colour one can tell when a *Purpura* has changed from one diet to the other.

The shield-shaped common limpet *Patella* is also a prominent member of the lowest part of the tidal zone. At ebb tide it sits hugging the rock, always in exactly the same place; the brim of its shell is shaped to match irregularities in the rock, so that it is as if the two were moulded together. The strange thing is that at high water, and particularly at night, *Patella* leaves its home and embarks on long wanderings. The length of its journeys can be seen by the trail it leaves, as it rasps the green algae cover on its way. If we follow the trail, we see that this often forms an irregular circle, sometimes with a large sometimes with a small radius. But the surprising thing is that before the next low tide the snail, with unfailing accuracy, returns to exactly the same spot on the rock, where its shell fits exactly, and where it can shut itself in during low water. We call this strange instinct homing, but we are not able to explain it. A *Patella* with its shell pressed tightly against the rock is almost impossible to remove. Rats hunting among the rocks sometimes try to prise the limpet off the rock with the teeth of their lower jaws, but this is a risky undertaking and occasionally they are trapped and drowned when the tide rises.

7·8 A whelk kills a mussel by prising open its valves, then tearing at it with the teeth on the tip of its proboscis. This method does not, however, succeed with the scallop (below), which uses its shells like bellows to push the invader away in a jet of water. (After G. Thorson 1968.)

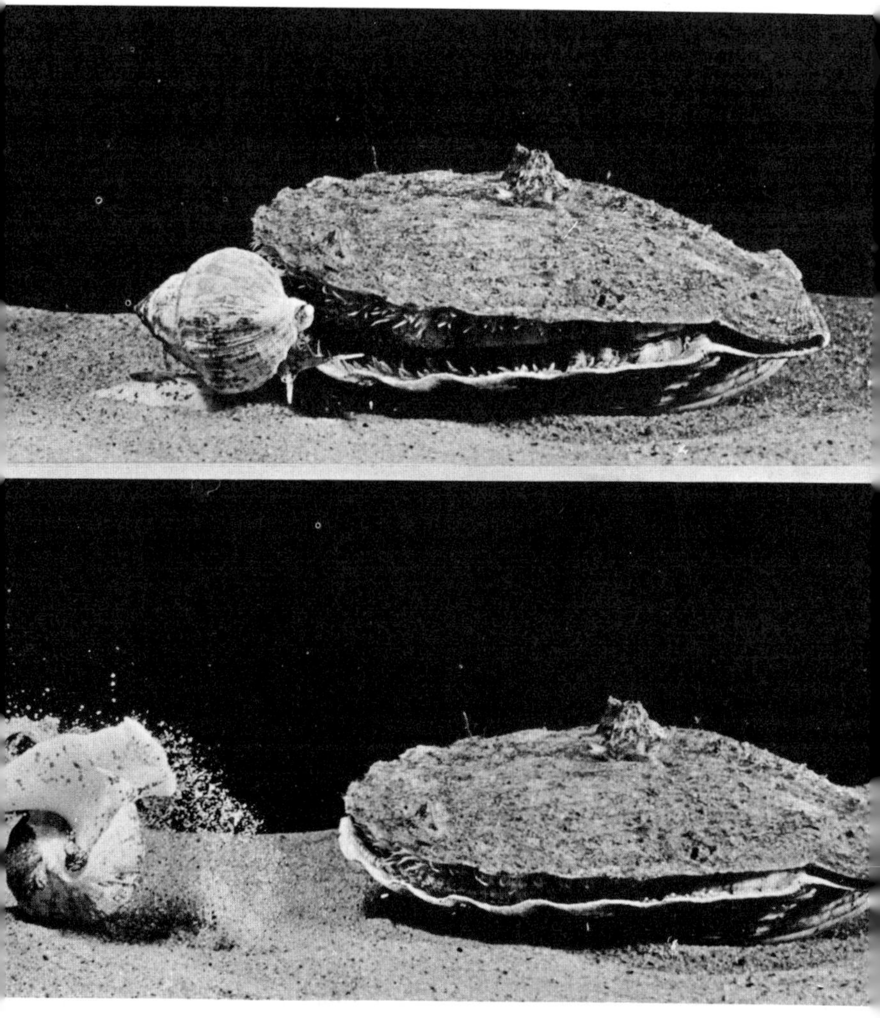

7·9 Dangerous and venomous inhabitants of a Pacific coral reef. Near the surface are a Portuguese man-of-war, a large jelly fish, and a group of barracudas (*Sphyraena*), a killer clam, *Tridacna* and, around it, stinging coral (*Millepora*). In the centre are the dragon fish *Pterois*, two electric rays, *Torpedo* (on the bottom) and a moray eel, *Muraena*. At the bottom of the picture, on the staghorn corals (*Acropora*), are a stone fish (*Synanceja*), a poisonous cone shell (*Conus*), a bristle worm (*Eurythoë*) and the sea urchin *Diadema*. (Kai Olsen.)

We have mentioned the common mussel (*Mytilus*) as a victim of the *Purpura* snail. All over the world mussels are the most characteristic animals of the tidal zone. They often occur in immense numbers – a mussel population weighing 30–50 kilograms per square metre is no rarity on suitable rock surfaces. In the Russian Arctic tidal zones the common mussel often makes up more than 70 per cent of the total animal weight per square metre; in Britain we know of recently established mussel populations with a density of 140,000 to 170,000 per square metre. The English zoologist D.A.Hancock has recently found that Danish mussels, although belonging to the same species as the English ones, are four times stronger than the British variety; they can close their shells more tightly and are more difficult for sea stars to open and, accordingly, less predated by these. Among the mussels live sea stars, sea anemones, many species of crabs and a multitude of other animals.

Coral and other reefs Among tidal areas dominated by epifauna we have not yet discussed the most luxuriant of them all – the coral reefs – scattered over 190 million kilometres in the tropics. The best known area is the Great Barrier Reef, which stretches for over 2,000 kilometres along the east coast of Australia from New Guinea to Queensland. The coral reefs are limited to a belt from approximately 30°N to 30°S latitude. They are therefore exclusively confined to the warm sea areas and even in these latitudes they are as good as absent on the west coast of the continent where an offshore wind blows the top-most surface layers of the water out to sea causing the cold bottom water to well up. Coral needs warm water; its optimum temperatures are around 23–25°C and below a temperature of 18°C large coral reefs are unable to form.

Another important factor in the formation of coral reefs is that

the water should be clear, that is, free from suspended matter which would collect on the coral and hinder its development. Clear water is important for another reason; coral depends a great deal on light and rarely develops in depths greater than forty metres. The need for light is linked with an interesting feature of the coral's biology – its symbiosis with a group of yellow-brown microscopic algae, the so-called zooxanthellae, which are actually vegetative stages of the dinoflagellate genus *Gymnodinium*. These unicellular plants live inside the cells of coral animals and occur in such numbers that their colouring often dominates the coral animals' own colour. All reef corals carry zooxanthellae in such numbers that there may be as many as 30,000 specimens per cubic millimetre of coral tissue. By the time coral larvae leave the mother coral and swim freely in the water, before starting a new colony, zooxanthellae are already present in their cells as a kind of imprisoned phytoplankton. It hardly seems likely that such a close connection would be worthwhile unless it were of value to both participants. In the case of zooxanthellae the advantage is obvious: they live in a sheltered milieu inside the coral cells and can directly utilise the natural waste products, not only carbon-dioxide, but also many other inorganic products such as nitrate and phosphate, for their own nutriment. In return the coral is spared the necessity of having to dispose of these waste products, and during photosynthesis the algae produce an extra supply of oxygen for the coral to breathe. Thus, mutual advantage is clear enough, though not all the points have yet been clarified. It is still being argued whether coral actually eats the zooxanthellae. C. M. Yonge, leader of the British expedition to the Great Barrier Reef, made a series of physiological experiments which seem to show that this does not occur. By keeping coral in the dark, where algae cannot survive and will soon be extruded by the polyps, he discovered that coral can very well live without zooxanthellae. He also found that individual coral polyps, just like sea anemones, are specialised carnivores, able to capture zooplankton and digest it with amazing speed. Yonge is also of the opinion that coral animals are not able to digest cellulose and, therefore, cannot feed on any dinoflagellates, including zooxanthellae. Some new observations however seem to show that the zooxanthellae may have a general stimulating action on the metabolism of the coral host, while the algae do not contribute significantly to the nutrition of the coral. The presence of the algae, however, seems to stimulate the calcium metabolism of the coral; corals with zooxanthellae grow appreciably faster than those without. It should be added that zooxanthellae are also found in symbiosis with many other animal groups: hydrozoans, medusae, sea anemones, sponges, flat worms, and even in the mantle of the giant clam *Tridacna*.

A coral reef often consists of much more than just coral, although in the enormous reefs in the Pacific, coral itself does dominate. The number of 'true' coral species is, however, much richer in the Indo-Pacific Ocean than in the tropical parts of the Atlantic. The reef coral fauna of the Atlantic comprises only about 35 species of 26 genera; that of the Indo-Pacific some 700 species of about 80 genera.

Most important in both areas are the genera *Acropora* (three Atlantic and about 150 Indo-Pacific species) and *Porites* (three and thirty species respectively). The same species can occur in very different sizes, depending on the depth in which it grows, the water movement it is exposed to, and the amount of suspended matter covering it. In calm water it can become as branchy as the finest tree; in a surf area it tends to be concentrated into a large solid mass. It is the inner skeleton of calcium carbonate left when the organic tissue dies off which actually builds up the immense reefs, but while young coral is building a new skeleton, many animals of the reef are busily destroying what the earlier coral has formed. Several species of boring bivalves, polychaete worms and sponges live in the calcareous skeleton of the corals, and as these break down the lime it is often

7·10 Staghorn corals of the genus *Acropora* are the dominant animals in most Pacific coral reefs. (After C. M. Yonge 1932.)

eaten by the legions of sea cucumbers that are associated with reefs. These can consume from 10 to 80 kilograms of detritus per animal per year. When the lime has passed through their digestive tracts, it has been transformed to a powder, which will sooner or later be deposited as a dense sediment on the sea bottom round the reef.

Fishes living along and upon the reef also contribute to the destruction of the corals. Some of the fishes are plant eaters while others feed on sponges. Grazing upon the corals, they bite off small pieces of limestone. The American geologist K. O. Emery found that fishes do as much destruction as is produced by the erosive effect of waves and currents.

If a reef is to increase its area, the growth of new calcareous skeleton must be greater than the destruction of the old, and this is by no means always the case. We know of reefs where the destruction is greater, and where, year after year, we can see the reef become progressively smaller. The growth rate of coral skeleton is not particularly impressive. The branched form of *Acropora* shows a skeleton increase of $2\frac{1}{2}$–5 centimetres per branch per year, but many other coral species will grow less than 1 centimetre. Under really favourable conditions a permanently submerged reef can grow towards the surface at a rate of about 30 centimetres in eleven years. This means it would take roughly a thousand years to attain a height of 28 metres. These, however, are only average estimates. We also know that growth rate declines with the increasing size and age of the coral colonies. For one genus, *Manicina,* it has even been found that the calcification rate of the smallest colonies was more than seventy times that in the largest. There are coral species which may grow up to a diameter of two to three metres, and others which will attain a height of only 35 to 50 millimetres as 'adults'.

There are many other organisms that help coral to form reefs. Large reefs, particularly in the West Indies, although known as coral reefs, consist of hydro-corals of the genus *Millepora*, the

7·11 The killer clam *Tridacna derasa* often reaches a length of nearly one metre and a weight of more than one hundred kilograms. (After C. M. Yonge 1932.)

so-called 'stinging coral' or 'fire coral'. It is a close relative of the hydroids but like the true corals its skeleton is mainly built of calcium carbonate which is far more massive than that of the coral reef proper. The colony is divided into a large number of mouths, each surrounded by a circle of threads with stinging cells. The sting can be very unpleasant, even for humans. *Millepora* reefs, like coral reefs, take many different forms. Sometimes individual colonies are strongly branched; now and again they are in solid lumps or form large, even surfaces, and it is difficult to believe that these reefs, so widely differing in appearance and found throughout the tropical zones, are all constructed by what many biologists regard as the same animal species, *Millepora alcicornis*. There is no doubt that *Millepora* is a polyp. It even produces short-lived free-swimming medusae, the larvae of which attach themselves permanently and form the polyps of the reef. Like the polyps these microscopic larvae have powerful stinging cells, and when they swarm in the water in large numbers they can inflict severe pain on bathers. Some years ago, in Miami, Florida, such larvae became a threat to tourism. It took biologists some time to discover that the stings were caused by the larvae of hydrozoans, almost certainly those of the fire coral, and it was also found that the distribution of larvae had pronounced maxima and minima – probably a moon phase phenomenon. By taking daily plankton samples, it was possible to detect dense concentrations and to warn bathers.

Among the reef builders other than the true corals, the bristle worm genus *Sabellaria* should be mentioned. These worms build individual tubes of grains of sand glued tightly together, and since the worms live intertidally in enormous aggregations, huge layers of tubes usually arranged in parallel rows one upon the other often form solid structures of one metre width or more. Such *Sabellaria* reefs occur in their most luxuriant form on the coasts of Brazil and South Africa.

Even algae take a hand in the formation of reefs. The most active algae are a lime-encrusted species of *Lithothamnium*. They make a red coating over the already existing lime formations, and some of them may directly grow out to form 'miniature corals' of only a few centimetres in height.

There are several different types of reef, there is the *fringe* reef, that is, a reef situated on the coast itself and forming a border along its outer edge. Then there is the *coastal* reef, or *barrier* reef. These are always situated away from the coast and between them and the coast there is usually a canal of deep water through which even large ships can pass. A third type is the *atoll*. This is the type associated with the idyllic Palm-tree islands of the Pacific. They consist of a ring-shaped reef, usually encircling a shallow basin – a lagoon – which is connected to the sea outside by an opening. A lagoon can have a diameter of more than 60 kilometres. While the first two

7·12 During the last few years the coral-eating crown-of-thorns sea star *Acanthaster planci* has increased enormously in number and is now a serious menace in the Pacific coral reefs. (After T. F. Goreau 1963.)

types of reef are bound to an existing coast, an atoll can emerge in the middle of the open sea, hundreds of kilometres from other land. These islands consist solely of coral limestone, and there are many theories as to their formation. However we must content ourselves with saying that the subject is still controversial and that no final explanation that would be universally accepted has yet been reached.

A coral reef has the most luxuriant and varied population of any area of the sea. It is impossible here to mention more than a few examples of the varied life it contains. However, the giant clam must be mentioned, partly because of its size, but also because it plays an important part in the Indo-Pacific coral reefs. *Tridacna derasa,* one of the seven or eight *Tridacna* species found in the reefs, is the largest mussel living on our earth today. There are specimens that measure 1·35 metres in length, 70 centimetres in width, and weigh about 200 kilograms. At ebb tide these giants lie completely or partly exposed among the coral. Despite their size they are hard to spot as the shells are often overgrown with algae and blend well with the surroundings. The soft parts and the voluminous mantle-edge in particular gleam with many shades of iridescent browns and greens – an excellent camouflage among all the other colours of the reef, and it is this camouflage that makes them so dangerous to human explorers. Divers who inadvertently place a leg or an arm in between the shells of the giant clam have very little chance of getting out alive, especially if they are already submerged. The shells close so tightly that the limb is trapped as if in a vice and it is very difficult to sever the clam's adductor muscle. Even on the coastal reefs cases have been known where tourists wandering about at ebb tide found themselves trapped by a giant clam and were unable to free themselves before the tide came in, with fatal results. As a consequence, these giant clams are also known as killer clams. Some species, when touched or irritated on the border of the mantle, spurt out a hard jet of water aimed directly at the intruder.

Another typical inhabitant of coral reefs is the black sea urchin, *Diadema*. The two predominating reef species, *D. antillarum* in the Atlantic and *D. setosa* in the Indo-Pacific are often found in quite shallow water in groups of up to a hundred or more, making the reef a 'danger zone'. This is because of their slender 20 to 30 centimetre long spines. If a hand or even a shadow is moved towards a *Diadema,* it immediately aims all its spines towards the 'attacker'. To be touched by a *Diadema* means pain for hours; the slender, fragile, but not poisonous spines penetrate deeply into the flesh, breaking off in the wound so that it is usually impossible to get them out. Many species of fish and shrimps seek shelter among these spines, but in spite of its defences, *Diadema* may be successfully attacked and eaten. About fifteen West Indian species of fish prey on *Diadema*, some of them carrying permanent scars with broken spines on their lips.

A few more characteristic reef animals should be mentioned. One

of them is the sea star, *Acanthaster*, the so-called 'crown-of-thorns star'. It is about 20 centimetres in diameter, violet-blue or orange in colour, and with 3 to 4 centimetre long poisonous spines. It is common in the Indo-Pacific, where it directly feeds on the corals, and has recently been found to be destroying large areas of coral growths of the Great Barrier Reef and other areas of the Pacific. Two genera of basket stars, *Astroboa* in the Indo-Pacific and *Astrophyton* in the Atlantic, have the five arms characteristic of echinoderms; but these branch over and over again to form an efficient and flexible trap. An adult *Astroboa* can filter some 800–900 cubic

7·13 The predatory Indo-Pacific basket star *Astroboa nuda* spreads its arms at night to form a trap which can filter up to 800–900 cubic metres of water in eight hours. (After H. W. Fricke 1966.)

metres of water in eight hours and for *Astrophyton* there are impressive figures of the number of prey it may capture in a night, including even small fishes. Finally, venomous fishes, such as the dragon fish *Pterois,* and the stone fish, *Synanceja horrida,* are also commonly seen on the coral reefs.

Actually a coral reef should be seen at night if one is to get a full idea of its teeming life. Most of its fishes seek shelter during the day, moving around only in the dark, as do the diademas. At night the feather stars and the basket stars unfold their fans and their nets, and the surface of the reef presents quite a different appearance.

Infauna of the littoral zone

We now move to those animals which live in the sand or mud of the sea bed (infauna), as contrasted with those that live mainly on rocks, stones, vegetation, etc. We are still in the tidal zone, and first in order below the steeper beaches come the sand flats.

Sand flats We have seen in the case of the jumping mollusc *Donax* that to live successfully in sand which is periodically disturbed by wave action an animal must be able to dig itself in very quickly. Once buried, an animal achieves protection from predators, and is free from the danger of being dried out at low water. It is a very specialised way of life, however, and it is not surprising to find that only a small number of species choose this habitat; but those species that do choose it are to be found in very great numbers. For example, the common European cockle, *Cardium edule,* is found in densities up to 350 per square metre, while in the Dutch Waddenzee two thousand specimens per square metre are to be found, each cockle being 25 millimetres in diameter.

Another inhabitant of the sandy bottom is a small 4 centimetre long bristle worm, *Thoracophelia mucronata*, living in the sand flats

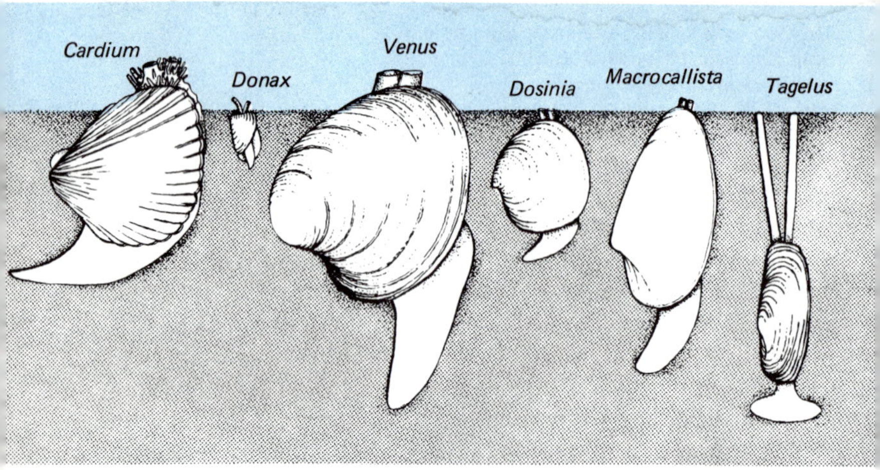

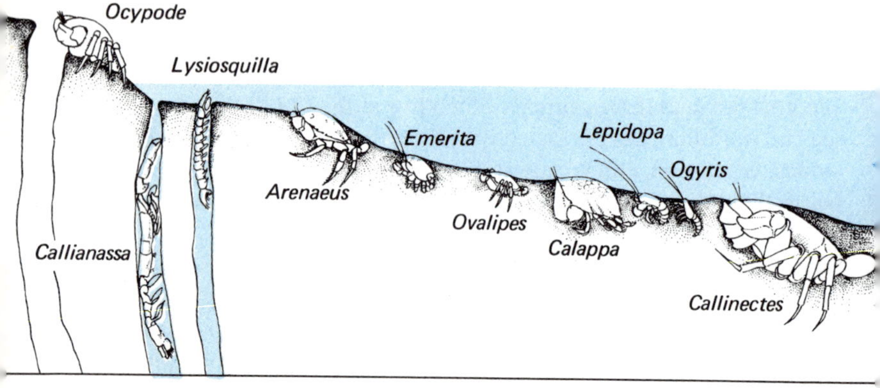

of southern California. This creature feeds itself by scooping large amounts of sand through the intestine and digesting the 1 per cent organic matter found in between the grains. About 3,200 of these worms are found per square metre, and they swallow sand at such a rate that a hundred worms are able to pass 210 kilograms of sand (2,100 times their own weight) through the intestine each year. This means in effect that if we take a sand beach, say $1\frac{1}{2}$ kilometres long and 3 metres wide, 14,600 tonnes of sand per year are 'filtered' through the intestinal canals of these worms.

7·14 Top A vertical section through a beach in North Carolina showing the burrows of bivalves in the transitional zone between sand and mud. (After A. S. Pearse 1942.)
7·15 Bottom A vertical section through a beach in North Carolina showing burrowing crustaceans in resting positions near the surface of the sand. (After A. S. Pearse, 1942.)

Another feature of the fauna of sandflats is that it consists either of large, sturdy forms which can bury themselves into or plough their way through the bottom, or of microforms so small that they can live in the interstices between individual grains of sand. Among the fastest diggers are the shrimp *Crangon,* the amphipod *Haustorius* (which has spadelike feet like a mole), and crabs of the *Emerita* type. Among the ploughing types are the cockle, the razor shell *Ensis* (which can fill the lowest part of its foot with liquid, causing it to bulge and form an anchor in the sand), and the tropical snail *Oliva,* with its enormous, wedge-shaped foot.

Mud flats This area of the tidal zone is almost horizontal. This means, among other things, that at low tide the water does not run completely away but remains in the mud. In this respect the fauna of the mud flats is in a more favourable position than that of the sand flats. Even at the lowest spring tide there is no need for an animal to dig very far down. Nevertheless, these animals have chosen a difficult environment. Only the topmost centimetre of mud contains any oxygen, and below this there are often layers with a high content of hydrogen sulphide. In addition, the mud particles are often so small that they clog the gill-filters of many animals. There is generally (but not invariably) more organic matter in a mud bottom than in a sand bottom, and this is the food on which the majority of mud-bottom animals depend. These are mainly detritus eaters, such as bristle worms with their long tentacles which comb the mud surface, mud-burrowing amphipods, *Corophium,* which sit in their tubes collecting detritus from the bottom or sieving it out of the water, and mud snails, *Hydrobia,* which actually eat the bottom surface layer with its content of bacteria.

Most of these mud dwellers are small, but there are a few exceptions, such as the massive bivalves which live on the west coast of the United States. The horse clam *Schizothaerus nuttalli*, reaches a

7·16 Right The Indo-Pacific sand crab *Matuta lunaris,* which lives in shallow water and can dig itself into the sand in only a few seconds.
(After E. C. Pope and P. M. McDonald.)
7·17 Far right The Indo-Pacific giant beach worm *Onuphis teres,* which can burrow through the sand with extreme rapidity.
(After E. C. Pope and P. M. McDonald.)
7·18 Bottom At low water, tracks of the snail *Olivella columellaris* cover the beaches along the Pacific coast of El Salvador.
(After A. Seilacher 1959.)

length of 20 centimetres and a weight of 2 kilograms, while the geoduck, *Panope generosa*, is even bigger with a maximum weight of 5–6 kilograms. It sits nearly 2 metres deep in the mud and probably lives as long as twenty years. They are both difficult to dig out, but it is worth the effort as they are exceptionally good to eat.

Mangrove swamps A special type of mud flat is to be found in the tropics covering long stretches of the coasts of Indonesia, Africa and South and Central America. These mud flats are traversed by canals and carry a dense growth of mangrove scrub, among which *Rhizophora mangle* appears to dominate in all parts of the world. Because of the shadow of its abundant leaves the mud bottom is at least partly protected from the tropical sun, and the mangroves can thrive both in brackish water (in delta areas) and in normal sea water. The mangrove swamp is a steamy milieu, a kind of transition area between sea and land, and many biologists believe that it was by way of areas such as these that the first sea animals began to conquer dry land about 350 million years ago.

Of the larger sea animals often found here in colossal numbers, the most conspicuous are oysters, which sit cemented to the immense root system of the mangroves, species of the ubiquitous common mussel, viviparous periwinkles found high up on the branches and stems, and an incredible multitude of crabs. The Brazilian mangrove swamps alone contain eleven known crab species, among which species of fiddler crabs, *Uca,* are both the most dominant and the most amusing. At ebb tide the mud flats of the mangrove swamps are alive with crabs. They often move in large groups and all in the same direction. The female resembles an ordinary crab, but one claw of the reddish-brown male is so exaggerated that it is almost as large

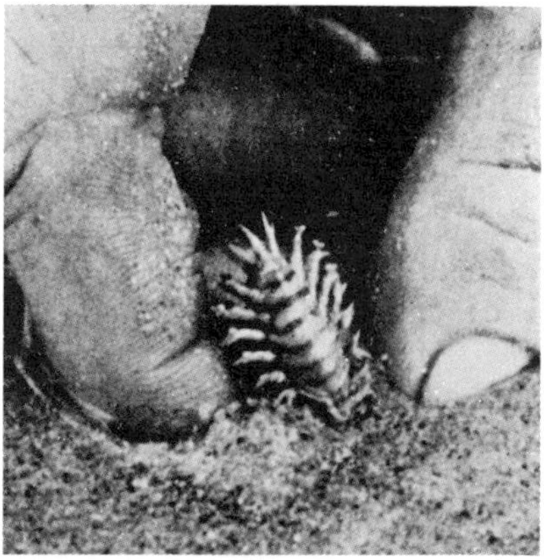
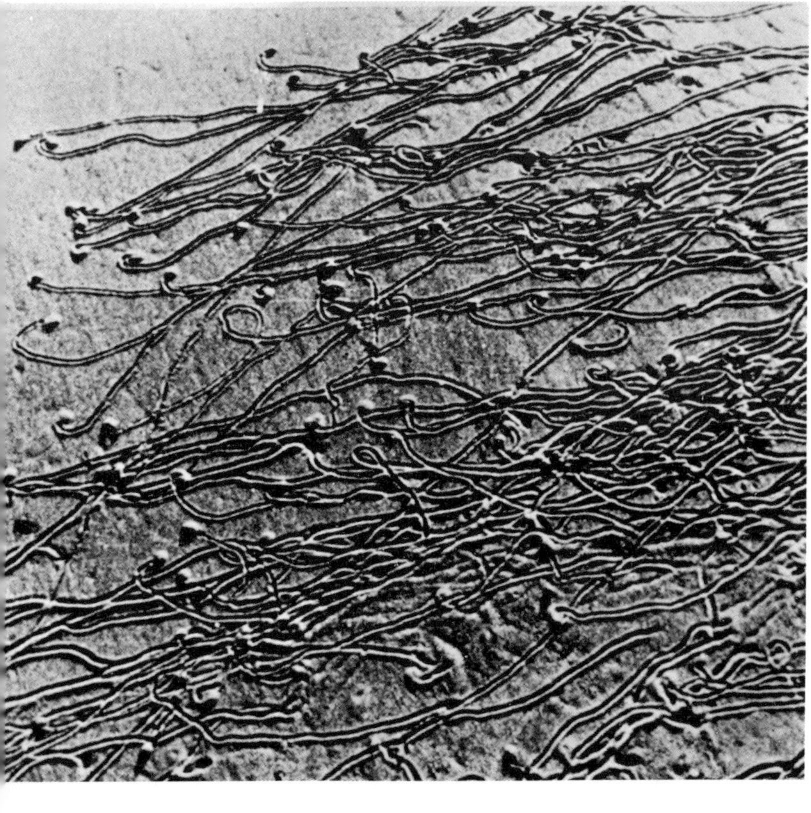

7·19 Below Fiddler crabs (*Uca*) at ebb tide on a Brazilian mud flat. The males have a single large claw (here one can be seen emerging from its burrow). In the middle is a female. (After E. E. Boschi 1964.)
7·20 Right The small crabs *Pinnixa rathbuni* can occur in densities of up to 3,000–4,000 individuals per square metre on sand-mud bottoms at fourteen metres depth off Japan. They dominate the community and feed on plankton. (After G. Thorson 1957.)

as the rest of the animal. This claw is a curious structure, yellowish-white in colour; with it the crab performs strange, signal-like movements just as though it were waving or playing the fiddle – hence the name. This large claw is used by the male for display in front of the female (in much the same way as the spread of a peacock), but it is also a ruthless weapon in mating fights. When the tide starts to rise all the crabs disappear into their tunnels in the sea bed, which are plugged with a piece of mud.

Another mangrove swamp-dweller is the mud skipper *Periophthalmus*. This is a small fish, 10–15 centimetres long, which spends half its life on land. Its eyes are almost mounted on stalks and are constructed to see all round both in the air and in water. The

10 cm

ventral and pectoral fins function as primitive legs, and when the animal is in motion the tail can be used to assist propulsion in the same way as that of a lizard. One sees it hop over the mud flats at a surprising rate, spring quite violently on to its prey, but then immediately disappear into the mud if it senses it is being followed. Although the mud skipper is the best known, there are many other species of fish which spend part of their life out of water. Prime examples are the blennies (*Blennius* and *Chasmodes*), which are usually about 10 centimetres long, with a long dorsal fin. Several species of blennies are seen almost as frequently on land as they are in water, and their technique of moving in an amphibian environment is very like that of *Periophthalmus*.

7·21 The mud skipper *Periophthalmus koelreuteri* lives in mangrove swamps. At the top is a diagram based on a motion film, showing its position when skimming over the surface of the water; at the bottom is shown the swimming posture, in dorsal and lateral views, when paddling (**a, c**) and when swimming quickly (**b, d**). (After V. A. Harris 1959.)

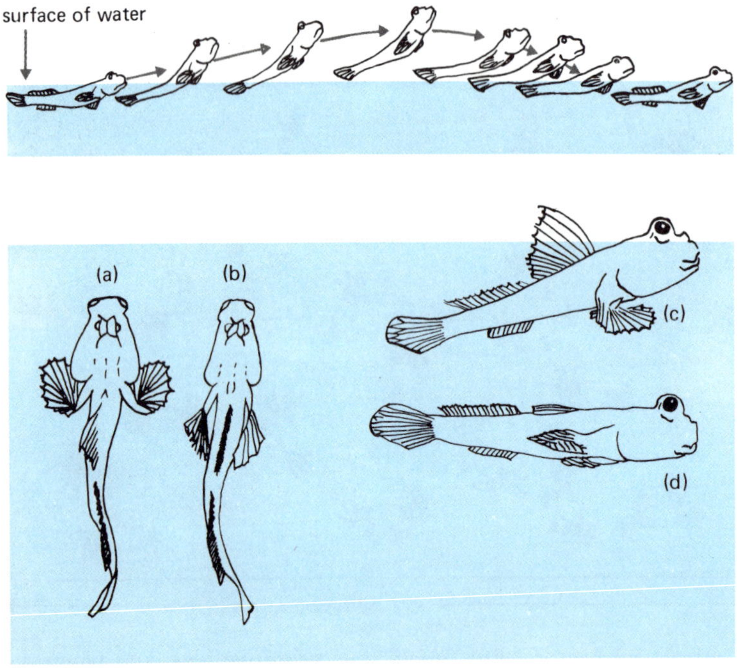

Depredation by birds At low tide the inhabitants of sand and mud flats are very accessible to birds and are eaten by them on a large scale. Many of the smaller wading birds eat enormous numbers of bristle worms and the amphipod *Corophium,* although it is difficult to ascertain exactly how many. However, in the case of the oyster catcher *Haematopus ostralegus,* we know that during winter in southern England this bird practically lives on cockles. Using binoculars, one can count the number of times it eats a cockle; and if a bank has been cleared of empty shells before the visit of a flock of oyster catchers, one can obtain a reliable total by counting the

7·22 A (stuffed) oyster catcher, *Haematopus ostralegus,* with its diurnal average consumption of cockles (*Cardium*), when feeding on a British bank in October. (After G. Thorson 1961.)

empty shells on the bank when the birds have gone. And the total is surprising. In October 1954 each bird on those mussel banks off southern England ate 214 cockles, each roughly 3 centimetres in diameter, every twenty-four hours. In January 1955, when it was colder and the birds needed more food to maintain their warm-blooded metabolism, they were eating 315 cockles a day. And an oyster catcher weighs only about 500 grams. Throughout the winter in question 30,000 foraging oyster catchers lived on the tidal flats, for a period of one hundred days in all. They thus ate a total of $30{,}000 \times 214 \times 100 = 642$ million cockles: this is calculated on the lower October figure and is therefore almost certainly an understatement.

There is no reason to believe that the small wading birds that usually feed on polychaete worms and on amphipods are less voracious. It has been calculated that the common sand piper, *Calidris alpina*, if feeding exclusively on 2–3 centimetre-long worms of the genus *Nereis,* will need about 450 worms a day, and we know that this worm is fully digested by a sand piper in twenty to twenty-five minutes.

It is worth mentioning in passing that large cockles sometimes act towards a wading bird in the same way as a giant clam does towards man. When the birds run over the mud flats they occasionally tread on an open cockle, which closes its shells with enough force to trap a toe or to cut it off. If we examine a flock of wading birds we often find that one or more of the distal segments of the toes are missing.

8 The sublittoral zone

The region from low water mark to about 200 metres depth comprises only a small area along the coast of the open ocean and corresponds roughly to the limit of the Continental Shelf. This *sublittoral* zone also more or less corresponds to the epipelagial of the free water masses: a zone where the surface light is still very important.

Borers and encrusters

Two elements of the epifauna are to be found at the surface, in the shape of incrustations on ships, and wood borers. These animals merit brief discussion since they directly affect man. Even a small amount of fouling reduces the speed of a ship and means loss of time and money. Large sums of money and much ingenuity are spent in trying to perfect anti-fouling paints with chemical properties which will effectively control these growths. In fact, more effort has been put into discovering preventatives than has been expended in examining the biology of the organisms responsible. These are algae, polyps, and above all, barnacles and calcareous-tube bristle worms; they are all fast growing forms which in the course of a few weeks can foul the hull of an otherwise well kept ship. A closer knowledge of the biology of these animals will obviously make the problem much easier to tackle. We know, for example, that barnacle larvae are extremely sensitive to the colour, structure and angle of light of the surface to which they attach themselves. By choosing the right colours on the right parts of a vessel one could immediately help to reduce infestation. As a last resort poisonous paints could be used, but the poison would have to be different for each type of organism. During the last few years, however, promising results in fighting fouling organisms on ships have been obtained by exposing them to ultrasound.

It should, however, be stressed that the number of species which

may settle on the hulls of ships and contribute to their fouling is relatively small. Most of the sedentary marine shallow-water forms which normally settle on rocks, stones or vegetation, are very specific in their choice of a substrate and are not usually found on ships. Thus, out of more than 250 species of barnacles known so far, only fifteen species at the most ever settle on ships' hulls, and the importance of ship-transport in the global dispersal of the marine shallow-water fauna is probably smaller than has hitherto been believed.

Boring animals are also a great problem. Here we must distinguish between two different types of marine borers: those boring in inorganic material and those boring in wood. The former group contains all varieties from organisms boring in stiff clay and mudstone through borers in sandstone and limestone to the few specialists capable of boring even in granite or gneiss. The sponge *Cliona* and flat worm *Pseudostylacus* mostly inhabit the shells of

8·1 Sea squirts often contribute to the fouling of the hulls of ships. *Ciona intestinalis,* shown here, can grow up to fifteen centimetres during a few summer months. (After G. Thorson 1968.)

living bivalves such as oysters and seem to use acid. The bristle worm *Polydora* seems to drill mainly with its laterally projecting spines and its jaws; the boring barnacle *Lithotrya* penetrates limestone by mechanical movements of some 'nail-like' bodies on its stalk; the seven groups of boring bivalves usually dig out their holes by sawing or filing movements of their fortified shells. The bivalves are able to bore in limestone but not in granite or gneiss, but we actually find some species of sea urchins which can drill even in this incredibly hard material, their five powerful teeth acting like a chisel against the individual crystals. Many boring animals start quite young with a small hole which is widened and deepened as they grow. They are automatically imprisoned in their hole and may have to tunnel for dear life; they can, however, still obtain food from outside and they have at the same time achieved an ideal protection against waves and enemies. Drilling sponges and worms can cause great destruction to oyster beds by boring through the shells of the living bivalves, and the boring bivalve called the piddock (*Pholas*) can work into a limestone coastline so vigorously that it is gradually eaten away.

It is the organisms that specialise in wood boring that are the most costly to man, chief among these being the ship worms (mainly the genus *Teredo*). Ship worms are bivalves but they differ greatly from the usual mussel type. They are about 15 centimetres long, with a yellowish-white worm-like body, of which only the foremost centimetre is covered by two short white shells. Their larvae, which resemble other bivalve larvae, swim around freely in the water, but when the time for permanent attachment approaches they fasten themselves to the first piece of wood they find. They already possess enzymes, among them cellulase, which can dissolve and digest the wood fibres (which are anyway usually already partly broken down by the activity of marine fungi). This helps the larvae to push far enough into the surface to be hidden from view, and here the final metamorphosis takes place. The shells are now reinforced with solid calcareous deposits and acquire small saw teeth along the leading edges, which are used for digging into the wood. The young ship worm swallows and digests the sawdust as it is rasped loose. It had been proved that 90 per cent of the cellulose contained in the ingested wood will disappear during the passage through the digestive tract of the ship worm; but it must also be remembered that the siphons of *Teredo* are in free connection with the surrounding water masses and by this means may add phytoplankton to their diet. The hole through which the small larva penetrates into the wood, and through which the ship worm, now imprisoned for life, still stretches out its siphons, is quite small, and gives little indication of the size to which the worms have grown inside the wood or the immensity of the internal tunnel system which is lined with a calcareous layer. A beam of a jetty which appears to be only slightly attacked can be almost hollow inside.

There is no doubt that depredation by ship worms the world over costs millions of dollars every year, and throughout the ages has led to a long series of catastrophes. Among the worst of these was an extensive attack in Holland, when in 1730 dykes and lock gates collapsed and had to be repaired at a cost of 8,000,000 guilders – a lot of money in those days. And in San Francisco, in 1857, a large number of houses built on poles along the coast subsided, completely undermined by teredos. There are modern biologists who believe that the much-debated defeat of King Philip II's Spanish Armada in 1588 may partly be explained by heavy attacks of ship worms. During the months when the Armada was assembled in Lisbon harbour preparing for the invasion, a larval infection from ship to ship may easily have taken place. Lisbon harbour offered an almost perfect basin for hatching teredos, and a few months would have been enough for the ship worms to grow and multiply with devastating effect.

8·2 The red whelk *Neptunea antiqua* is the largest snail in Scandinavian waters, and often has boring sponges (*Cliona*) mining in its shell. Only the oscula of the sponges can be seen, appearing as small yellow points on the surface. In this case, a young dead-man's-finger coral (*Alcyonium*) with its expanded polyps, is transported by the snail. (After G. Thorson 1968.)

8·3 Distribution of algae and epifauna on a fairly sheltered rocky shore in England. (After J. R. Lewis 1964.)

Another wood borer, the gribble, an isopod of the genus *Limnoria* measuring 4 to 5 millimetres, also plays a part in attacking wooden harbour installations. This animal also eats wood fibre but its main nutriment is the network of vegetable fungi, filaments of which are always woven into timber which has been immersed some time in water. *Limnoria,* often called 'the termite of the sea', is far from being as dangerous as the ship worm; it only attacks wood exteriorly and as the damage is detectable in good time suitable counter-measures can be taken. Nevertheless, the cost of its attacks can add up to millions of dollars a year.

Epifauna and algae

The epifauna in the sublittoral zone is closely connected to the epiflora – the permanently attached algae. An abundant algal growth is found on practically all rocks, stones or other hard substrates in the photic zone. On the narrow shelf around the Orkney Islands alone the wet weight of these brown algae is calculated to be around one million tonnes, and along the coasts of the Barents Sea other species of brown algae, *Ascophyllum* and *Fucus,* may in their optimal zone produce an average weight of about 15 kilograms per square metre.

But here, as in so many things, the United States can outdo anything we are able to offer in Europe. Along the Pacific coast algae grow to a length of almost fifty metres. The largest growths are found in depths of 8–30 metres, and the algae stand in the water like creepers in a tropical forest. In one species they are supported by floating bladders nearly the size of a football. In places the algae form a belt five kilometres wide along the coast of California. *Nereocystis* is the one family of kelp algae that is an animal, but all the same it can grow to thirty metres in a year. The other genus, *Macrocystis*, is perennial, and grows even longer. During the two

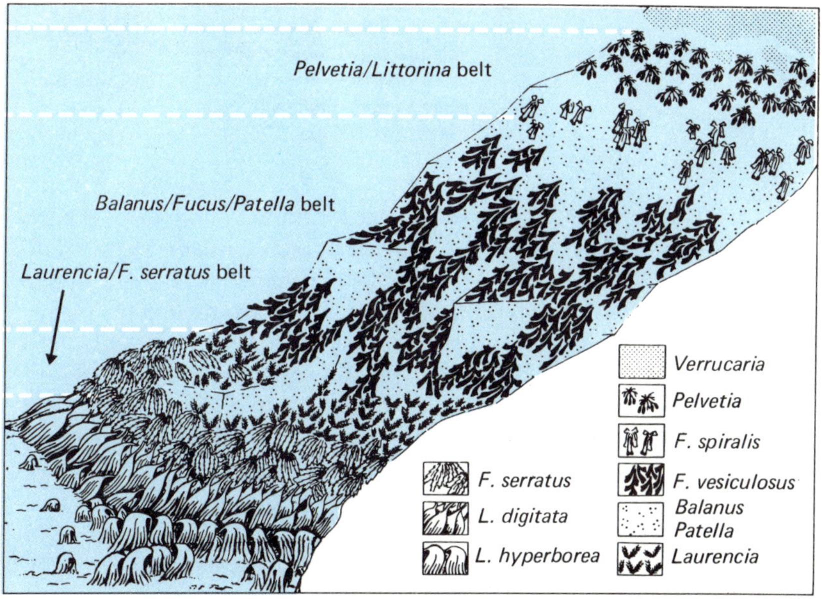

world wars attempts were made to exploit these algal forests on a commercial basis, not so much for food but for their content of important chemicals. For example, algae contain a concentration of iodine 20,000–30,000 times larger than that of the surrounding water, and a long list of other chemicals may be found in the algae in sufficient quantities to make exploitation feasible. Special boats were constructed for harvesting the largest algae, with large chutes mounted on the bows of each boat, extending obliquely to the required depth, and the bottom of each chute fitted with a power driven cutter-bar. As the algae were cut they were brought into the boat on a conveyor belt which occupied the full width of the chute. The forests of *Nereocystis* grew so quickly that they could be harvested four times a year. Even so, the procedure was barely economic and has in the long run never been followed up. It can, however, be taken for granted that the idea of harvesting and exploiting the large algae will recur again and again, until a technically practicable and profitable method has been developed.

It should be mentioned that, in the Philippines and especially in

8·4 Diagrammatic cross-section of a *Nereocystis* kelp bed near Carmel, California. Depths are in metres below mean low water. The horizontal scale is approximately the same. (After J. H. McLean 1962.)

Japan, algae are used on a large scale for human food. Genera such as *Porphyra, Acanthophyra, Caulerpa* and *Codium* are directly cultivated for food and some of them together with those of the genus *Furcellaria* in Scandinavian waters are also used in the manufacture of agar.

There is, however, one more way in which the forests of coastal macro-algae along the coasts may be important. Dense forests of the large algae act as a most efficient protective belt along many shores. Waves and currents may be subdued; mussel and oyster beds can develop in relatively calm water, and so on. Realising this, the Americans have already started experiments with artificial algal belts. Dense aggregations of broad plastic strips standing vertically in the water act much in the same way as a natural bed of *Laminaria,* thus giving shelter for beaches which have previously been exposed. The sedentary macro-algae can also serve as a substratum for the luxuriant growth of other plants. *Cystoseira barbata,* a medium-sized brown alga common along the shores of the Black Sea, is sometimes so densely covered with crusts of epiphytic diatoms that these amount to two-thirds of *Cystoseira's* own weight.

Living on the European *Laminaria* meadows, as well as the Pacific algal forests, are a large number of animals of many different groups. But there is one characteristic thing about these algae: in their living adult state they are eaten by comparatively few animals. When we look at trees and bushes on land, we see that their leaves, especially in late summer, are often badly damaged by insects and other animals that eat them. One sees surprisingly few signs of such attacks on the large algae. Most species that are epifaunal on algae live on the smaller algal tufts or on the colonies of polyzoa or hydroids which form growths on the large algae. There are exceptions, however, and one of these is the sea hare, an opisthobranch of the genus *Aplysia.* The Californian *Aplysia* species can reach 25 centimetres in length and weigh several kilos. If one gives them a stem of *Nereocystis* the

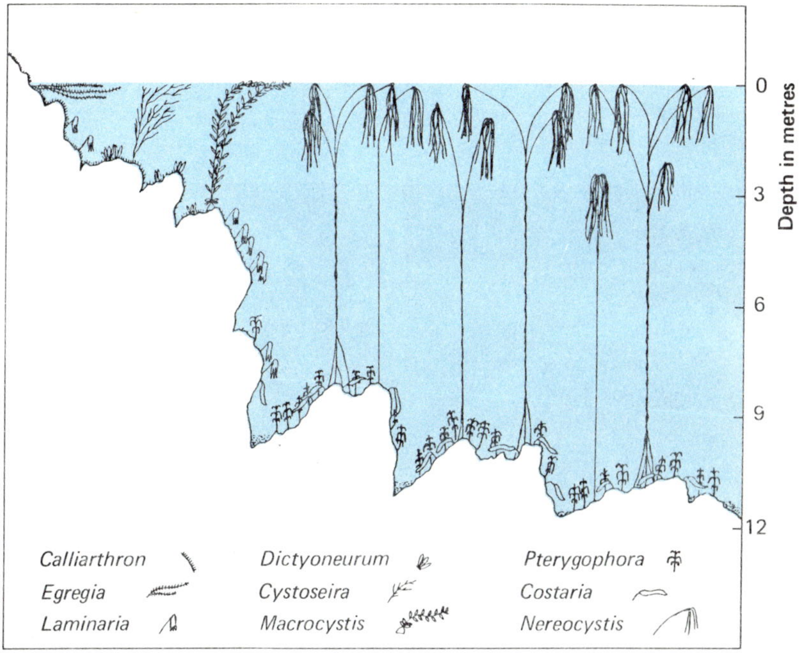

thickness of a broomstick one can hear their jaws crunch away at the crisp algae tissue, rather like a child chewing a raw carrot; half a metre of stem is probably regarded as a reasonable meal every twenty-four hours. It has been shown that the *Aplysia* specimens change their body-colour in accordance with their diet. A 6–8 centimetre long sea hare from the Scandinavian seas prefers, in its younger stages, to feed on red algae and will turn the same colour as these. Growing older, it usually migrates to more shallow water, where it feeds on the brown algae, and very soon changes to a greenish-brown colour.

It must also be pointed out that the absence of the newly settled sprout of sedentary macro-algae can have a decisive effect on the suppression or growth of algal meadows or forests along the coasts. On the Isle of Man the following experiment was made: on a well-defined section of the rocky tidal zone between high and low water mark all specimens of the common limpet, *Patella vulgata,* were

8·5 A vertical air view of a foreshore on the Isle of Man, showing algal growth after removal of all limpets (*Patella*) on a strip from high to low water mark. The arrows denote the algal growth at periods from a few months to a year after removal of the limpets.
(From S. M. Lodge 1948.)

removed. This species is known to rasp away the algal carpet covering the rocks and also the new young shoots of macro-algae. One year later, a big change was evident; a photo taken from an aeroplane clearly showed how the section that had been stripped of limpets contrasted by its luxuriant algal growth with the surrounding rock sections, where limpets had been allowed to browse undisturbed. Similarly when the oil from the stranded tanker *Torrey Canyon* covered and polluted the rocky shores along the southern coasts of England, the limpets succumbed in huge numbers and the algal meadows increased to unforseen quantities.

Still more devastating to the young algal sprouts are sea urchins. Such genera as *Echinus, Paracentrotus* and *Strongylocentrotus* can completely strip the rocks of vegetation with their sharp teeth. A recent study has shown that a density of just one *Strongylocentrotus franciscanus* to ten *S. purpuratus* or ten *Lytechinus anamensis* per square metre attacking the anchorage and associated structures of the giant kelp *Macrocystis* is enough to prevent any vegetation from developing.

A large number of different species of sea slugs, often showing splendid, sometimes even neon-like colours, can be seen browsing upon the incrustations of moss, animals and hydroids, which cover the surfaces of the larger algae. Most of the slugs produce secretions, often acidic, which prevent fishes, for instance, from eating them. Some of them even produce secretions which produce blisters on the human skin.

Epifauna of caves We can also find localities with a rich epifauna which does not live on large algae, for example in the under-water caves explored by frogmen in recent years. Here the sponges (related to our bath sponges) are dominant, sometimes covering the walls with a thick incrustation. Polyzoa and hydrozoan colonies also achieve their greatest expansion here. Sea anemones, built like lovely corals, hang down from the rocks like flower beds. Many of them are voracious predators which pull any prey that touches their tentacles down into their bag-shaped stomachs. They can starve for years if necessary. Although sea anemones, with their stinging nettle cells, are mostly well protected against enemies, they are preyed upon by nudibranch slugs and by some sea stars and crabs. A common sea anemone in the Samoan Islands, *Actinia kraemeri,* is even used as human food.

Octopuses also prefer to live in caves. They often build up a wall of stones around the body and sit motionless and well camouflaged, waiting for a prey to turn up and ready to change their colour pattern at a moment's notice. The prey will be seized by the suckers on the arms and torn by the horny, parrot-like beak. In certain species the bite has been shown to be venomous, and from Darwin in northern Australia there are reliable data on octopus bites fatal to man. Nevertheless, most of the tales about the size that octopuses can attain and their danger to divers are much exaggerated.

However, there is good reason for keeping clear of another group of animals, the morays, which also prefer the twilight of caves. These are a powerful species of eel, about one metre in length; they are voracious predators with long, pointed teeth and as many of them have a poisonous bite they should be treated with great respect. Some species are very aggressive and seem to attack anything that provokes their combativeness. They are unknown in cold-temperate waters, but frogmen diving in the Mediterranean and the tropics have unpleasant stories to tell about them.

Other benthic animals But back to the cold-temperate latitudes – to the large aggregations of the common mussel, forming epifaunas in only a few metres of water. Common mussels (*Mytilus*) are among the fast growing forms; our common North Atlantic species, *Mytilus edulis,* can reach full growth in eighteen months to three

8·6 The predatory sea anemone *Tealia felina,* here using a bottle as a substratum. (After G. Thorson 1968.)

years, and a single female can produce up to twelve million eggs. No wonder this is a dominant family in practically all the shallow water areas of the world. Mussels make excellent food for humans and are one of the few species really worth cultivating. A bunch of twigs suspended in the Danish Limfjord to serve as a substratum for settling and growing mussels, can in two years yield forty kilograms of mussels of which twenty kilograms are already of marketable size. In the Gulf of Taranto, southern Italy, where extensive beds of mussels are commercially cultivated, up to twelve kilograms of mussels may be produced per square metre every year. Unfortunately, common mussel growths attract other animals; among these, sea stars (especially the quite young ones) are important predators.

Crabs also belong to this area. In our waters the common shore crab, also known as 'the green crab', *Carcinides maenas,* is particularly numerous. These are voracious predators that eat everything from small live gobies to mussels and small snails. In the Elsinore Laboratory we have seen a hungry adult crab open and eat six large cockles in twenty-four hours. As mollusc eaters they are particularly harmful, something seen to great effect off the Atlantic

8·7 Left The moray eel *Muraena*; it has long, pointed teeth and, in some species, a poisonous bite. (After W. M. Stephens 1963.)

8·8 Below An adult shore crab *Carcinides maenas,* here partially hidden in the sand of an aquarium after having opened and eaten six cockles (*Cardium*) in twenty-four hours. (After G. Thorson 1968.)

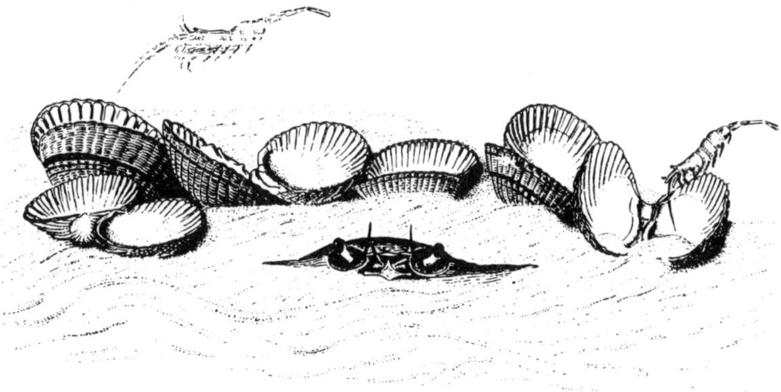

coast of Canada. They immigrated there fairly recently and have caused great depredation in the large soft-clam (*Mya*) populations which are of commercial interest to Canada. There are also the large edible crabs, *Cancer pagurus,* which, off the coast of England, live among the common mussel and are able to undertake migrations along the coast of at least 1,500 kilometres. Among the fish of the sublittoral zone, the wrasse, the stickleback, the viviparous blenny and the lump fish deserve mention. The lump fish lays its great egg mass of some hundred thousand eggs in spring at a depth of a few metres, where they are guarded by the male. These eggs, in the form of 'Danish caviar', have become a popular food.

At greater depths in the cold-temperate north-east Atlantic we meet the *Modiola* epifauna, which, as far as species are concerned, is the most luxuriant society these seas can offer. *Modiola* is a very large, slow-growing common mussel which lives for some twenty years. It is attached by a byssus to stones at 20–40 metres depth and is encrusted with colonies of the yellow 'dead men's fingers' coral, *Alcyonium,* and a multitude of brittle stars, the many-armed sea star or sun star, *Solaster,* sea anemones and strange crabs. The

8·9 This number of adult soft-shell clams (*Mya arenaria*) was dug out of the mud flat (shown at the top) of only one-tenth of a square metre on the German North Sea coast. (After H. Kühl 1951.)

8·10 A vertical section through a sand beach to show the depths at which the soft-shelled clam *Mya arenaria* is found at different ages and sizes. (Kai Olsen.)

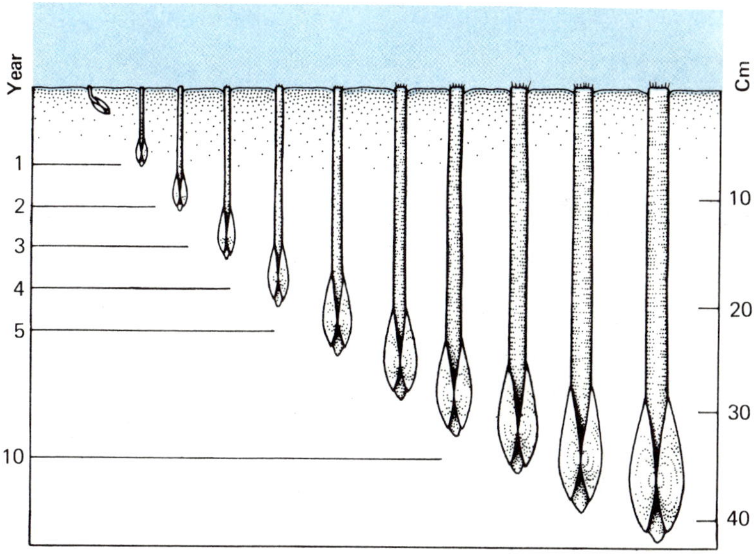

Modiola banks are very popular with cod fish and are usually good cod fishing grounds. Although the eggs of the cod, up to 4 million per female, float on the surface, and although they have pelagic larvae, the growing young spend the first two years foraging on the bottom. The older and larger cod usually feed on other fishes, but during the period of adolescence when they are most voracious, bottom animals, preferably crustaceans, are their favourite food.

People are always surprised to hear that there are also coral reefs in cold-temperate seas. At a depth of 100–200 metres from Bohuslän, western Sweden, and along the Norwegian coasts we find true reefs formed by the stony corals *Lophohelia* and *Amphihelia*. These do not stretch very far, and of course lie far too deep for the coral to cohabit with single-celled algae, as in the tropical reefs. Nevertheless,

8·11 A *Modiola* epifauna. The mussel to the left has a young colony of dead man's fingers (*Alcyonium*); a hydroid (*Tubularia*); two brittle stars (*Ophiothrix* and *Ophiopholis*); tube worms (*Pomatoceros*); and a barnacle on its shell. The mussel to the right carries a large hydroid (*Hydrallmania*) with a goose barnacle (*Scalpellum*). In the foreground are other hydroids (*Abietinaria*); a crab (*Eurynome*); a top shell (*Calliostoma*); red algae; and a colonial sea squirt (*Botryllus*). (After G. Thorson 1950.)

individual coral colonies, standing sometimes one metre high, can be beautifully branched. Among them lives a host of other animals: one oddity among these is the large basket star, *Gorgonocephalus,* with its great number of many-branched arms. Reefs of this type can be found as far south as Gibraltar and as far north as Lofoten. At great depths off the Scandinavian coasts we also find epifaunas dominated by brachiopods and sponges.

Sublittoral infaunas such as the *Macoma* community have 'parallels' in Arctic, cold-temperate and some warm-temperate seas, where other species, but of the same family, predominate on roughly the same kind of bottom and depth. Many readers will recognise the spiral excrements of the lug worm, when accumulations lie coiled on the bottom in clear water. Many of us have at some time

8·12 Vertical section through a *Macoma* community. (**a**) The tiny mud snail (*Hydrobia*); (**b**) a small whelk (*Nassarius*); (**c**) a cockle (*Cardium*); (**d**) a bristle worm (*Pygospio*) in its tube; (**e**) the common shrimp *Crangon*; (**f**) the bivalve *Macoma baltica*; and (**g**) the lug worm (*Arenicola*). (After G. Thorson 1968.)

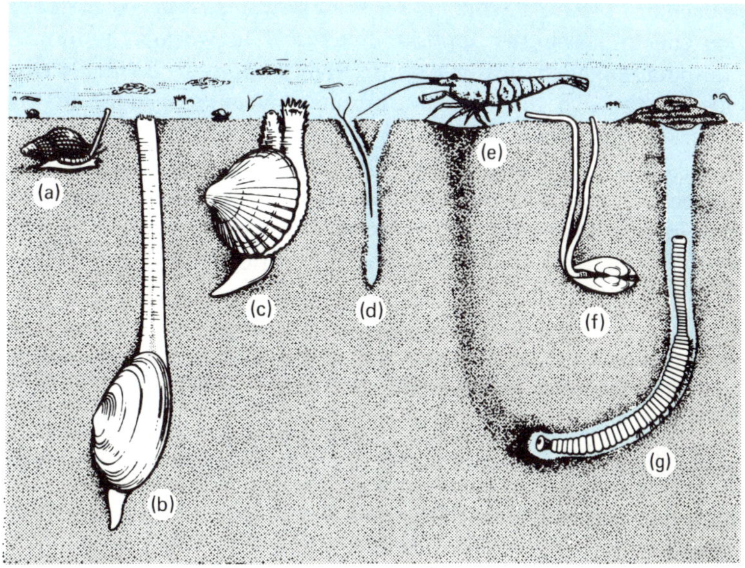

or another dug up these worms and used them as 'bait'. The lug worm *Arenicola marina* sits inside an almost L-shaped tube, the hind part of its body turned vertically upwards under the excrement coil, the forepart lying horizontally in the lower part of the L-tube. A funnel can be seen at the surface, above the worm's fore-end.

It is known that this worm eats sand in large quantities, and it was earlier thought that the funnel appeared because the animal sucked the light surface sand in a stream down to the fore-end of the tube and thereby to the mouth. It is true that the surface sand reaches down to the worm via this funnel, but the content of organic material on the surface is sometimes less than that found in the forepart of the worm's intestine. Roughly speaking, the lug worm swallows the sand, which sinks down through the funnel and is defecated at the

8.13 Vertical section through a sand beach and the depths at which the bivalve *Macoma baltica* is found at different ages and sizes. (From G. Thorson 1968.)

rear end of the tube, while its respiratory water is sucked in at the rear end to pass in the opposite direction. The worm's rhythmical pumping makes the water pass through its body to be finally discharged through the funnel. Carmine powder suspended in the respiratory water will be retained by the sand in front of the worm's head, and will soon after be found again in the fore gut of the worm. It seems reasonable to assume that all small planktonic organisms entering the tube with the respiratory water will suffer a similar fate. By this technique the worm may 'enrich' the sand before it swallows it. For long daily periods this may be its preferred way of feeding, while populations living close to high water mark, and out of water for long periods, may still behave as normal detritus feeders. It is known that even the sand without enrichment contains nutriment enough for the lug worm's demands.

Other families dominating this bottom are the cockle, *Cardium,* and the bristle worms *Nereis* and *Pygospio* (the latter is quite small but often occurs in such large numbers that its thin vertical sand tubes are as closely packed as organ pipes). All the community's animals, with *Macoma*-bivalves heading the list, are slow-growing forms which reach an age of several years. One often comes across large numbers of shrimps on the *Macoma* bottom: the families *Crangon* and *Leander* found in cold-temperate waters, and the family *Penaeus* in particular predominating in the tropics, where shrimps and prawns often seem to be much more numerous and also to grow to quite different sizes than they do in colder seas. For a zoologist wishing to find out how many animals there are per square metre of bottom, shrimps present an irritating problem as there is no way of catching them quantitatively. They rarely appear in our bottom samples, but as they seek their food among other bottom animals they are very much part of the community. From the German Waddenzee, figures are available to indicate that the number of *Crangon* eaten by local fish and harvested by the fishermen amounts

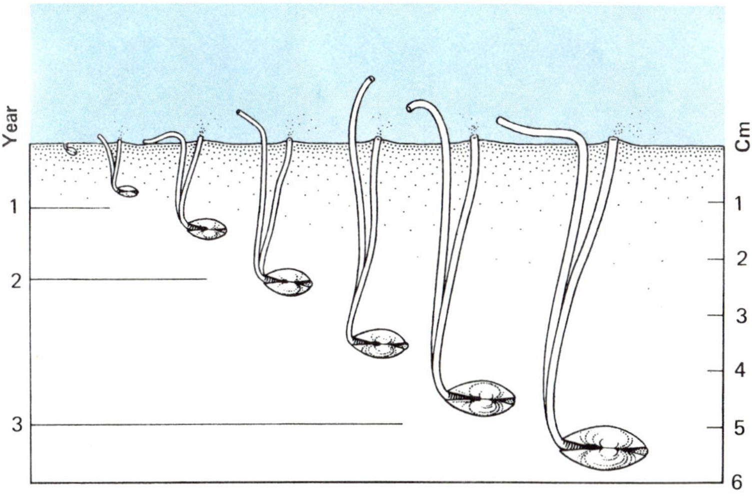

to about fifty specimens per square metre per year. The common small Scandinavian shrimp can eat everything from meiofauna to small gobies.

A *Venus* community can be found at depths of around 10–20 metres in pure sand in many seas – arctic, cold- or even warm-temperate. *Venus* is a thick-shelled, slow-growing mussel, but in places where the community is well developed there usually also occur species of the fast-growing, thin-shelled bivalve genera *Spisula* and *Mactra,* which are excellent fish food. *Spisula* in particular serves as food for the large, fast-growing plaice population of the Dogger Bank and the north-west Kattegat. At the same time, the boring snail *Natica* or *Polinices* eats a large proportion of the bivalves and is a strong competitor of the flat fish. With their sharp rasping radula teeth, and probably in most cases helped by an enzyme of the calcase type, they drill funnel-shaped holes through the shells of living bivalves, insert their probosces and ingest the soft parts. They are known to be voracious, and at least some species will on an average eat an amount of bivalve flesh corresponding to one-fourth of their own weight every day.

The same can also be said of the sea star, *Astropecten,* a predatory species which has been the subject of special studies at the Elsinore

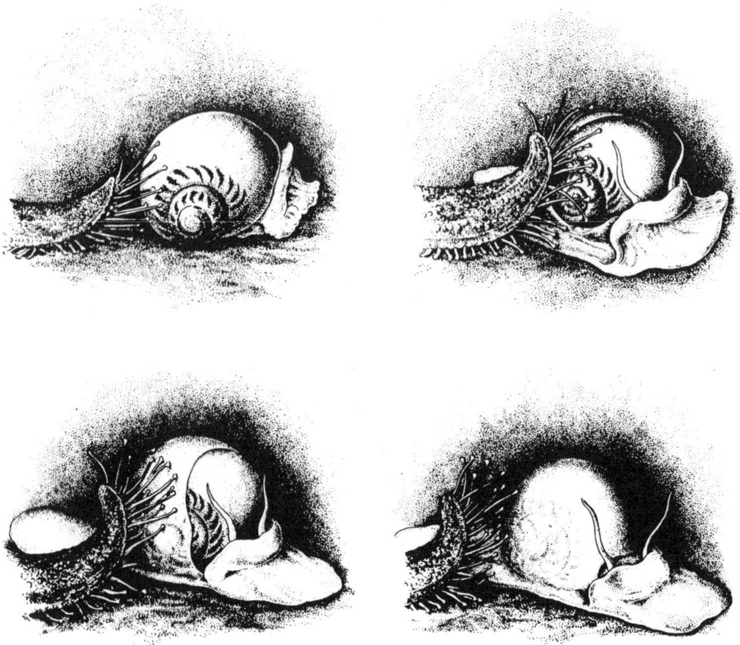

8·14 Top Holes drilled by the predatory snail *Natica nitida* through the shells of two bivalves and one of its own species. Similar holes are drilled by the closely related *Polinices.* (After Ziegelmeier 1954.)
8·15 Bottom The drilling snail *Natica catena,* when touched by the sea star *Asterias rubens,* swells by taking in water. A transverse fold of skin behind the shell is drawn forward like a hood (below left) to cover the shell. (Poul H. Winther.)

Laboratory by Aa.Møller Christensen. Unlike our common sea star *Asterias,* this animal is unable to turn its stomach inside out to envelop large prey but must swallow food in order to digest it. This species gives an instructive if surprising picture of 'the balance of nature'. If *Astropecten* were to eat *Venus* and *Spisula* on the same scale, this would soon lead to the decrease of the slow-growing, long living *Venus,* while *Spisula,* which grows to sexual maturity and full size in a year, would be better able to tolerate such 'taxation'. It now appears that *Astropecten* is able to swallow both *Venus* and *Spisula*, but their chances of surviving in its stomach are quite unequal. *Spisula* has a high metabolic rate and must frequently open its shells to breathe, whereupon the gastric wall of *Astropecten* is able to penetrate and digest it. *Venus* grows slowly, has a low metabolic rate, and can do without oxygen much longer than *Spisula*. By keeping its shell closed for a considerable period it is, therefore, able to withstand the gastric juices. The completely logical if surprising result is that a *Spisula* is digested in one or two days and the empty shell expelled, while a *Venus* can survive eighteen days in the stomach before being rejected – alive and kicking. Thus, with their quite different conditions of growth, these two mussels are able to maintain their populations on the bottom despite the attacks from *Astropecten*.

On mixed sand-mud bottoms in Denmark, Great Britain, the Mediterranean and the Black Sea there are a number of parallel communities living in the sheltered estuaries, among which mussels of the genus *Syndosmya* dominate. This is an oval, thin-shelled and fast-growing animal, and it occurs with a series of other fast-growing forms, including the bristle worms *Nephthys* and *Pectinaria,* of which the latter sits embedded in the bottom in an elaborate sand tube shaped like an elephant's tusk (widest end downwards). These communities are admirable places for the growth of flat fish. Particularly well known is the *Syndosmya* community of the Danish

8·16 When the bivalve *Spisula elliptica* scents a hunting *Astropecten* it pushes out its long, muscular foot and springs out of the bottom, landing six to eight centimetres away. (A. Margolin 1962.)

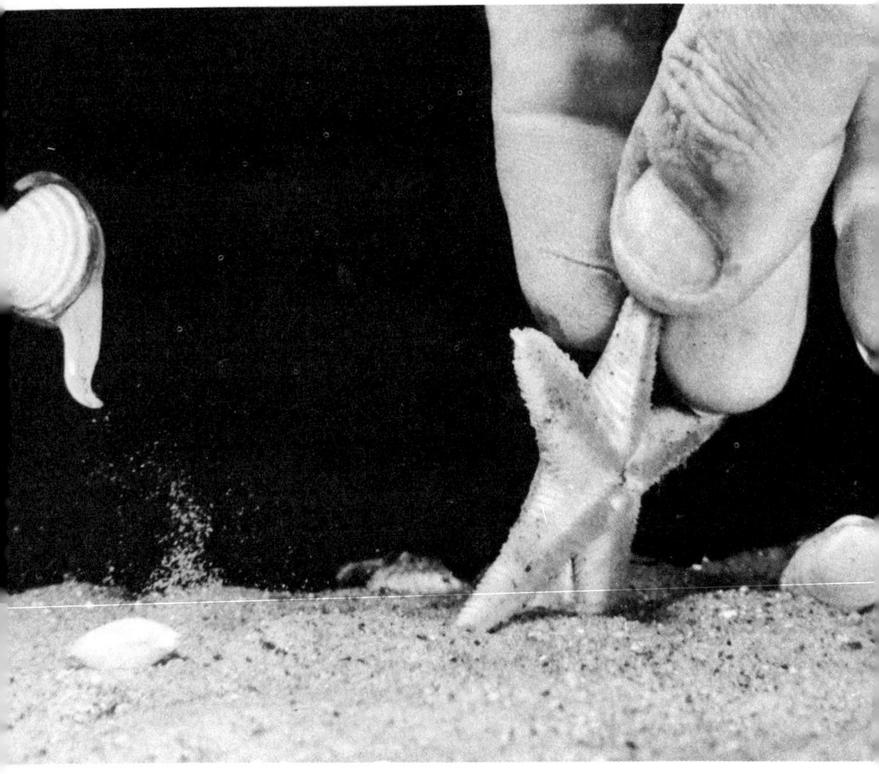

Limfjord, which Petersen and Boysen Jensen used for a number of investigations on production which are now regarded as classical.

Mud-bottom fauna from 10–20 metres to around 100 metres depth in Scandinavia, the Black Sea, West Africa, the Bering Sea, Japan, California, Chile and New Zealand are inhabited by thick layers of brittle stars of the *Amphiura-Amphiodia* type: echinoderms with

8·17 One-half of a square metre of a Danish mud-bottom *Amphiura* community, consisting of sixty specimens of the brittle star *Amphiura*; three large brown bivalves (*Arctica*); five heart urchins (*Echinocardium*); two sea pens (*Virgularia*, bottom of picture); ten tower shells (*Turritella*); one pelican's-foot shell (*Aporrhais*); and different small bivalves and bristle worms. (After C. G. Joh. Petersen 1918.)

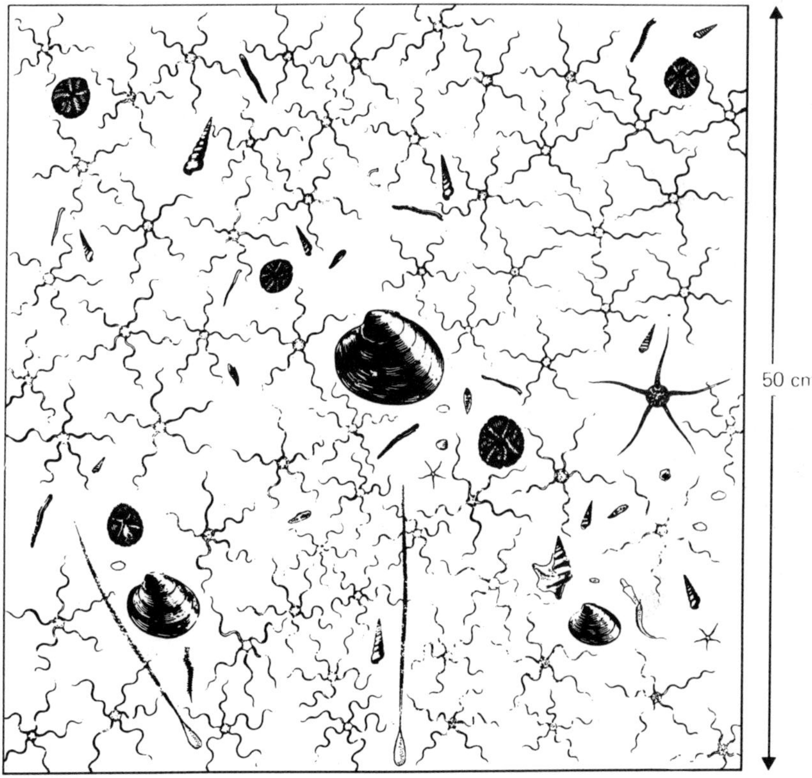

round soft disc and five thread-like, very supple arms. Considering that most people have never even seen a brittle star it may come as a surprise to learn that in the above-mentioned parallel communities, 400–500 of these forms per square metre are regularly distributed and that similar communities cover half the sea bed of the Kattegat, with a total weight of around 300,000 tonnes of brittle stars. Some

8·18 Like the heart urchin, the hermit crab *Eupagurus bernhardus* shovels large quantities of unsorted bottom material into its mouth. The hermit's smallest cheliped is used as a scoop. Indigestible material, having been rejected by the mouth parts, is deposited in a heap in front of the animal. (After G. Thorson 1968.)

brittle stars, for instance of the genus *Ophiura,* are fierce predators by nature but can maintain themselves for some time by digesting organic matter found between mud particles. In other words they can 'tick over' on a mud diet, but if meat extract is sprayed over a brittle star of the genus *Amphiura,* it becomes wildly excited.

The question arises of how it is possible for the larvae of the fish-food animals, and the enormous numbers of minute young mussels, snails, bristle worms, etc. (which leave the free water masses for the sea bed in late summer) to live among these brittle stars without being immediately discovered and eaten? I have personally seen that amphiurans can catch newly settled young of bivalves and worms which, sticking to the mucus of the brittle star's tube-feet, are hauled into the mouth. It seems reasonable to assume that such newly settled young together with the bottom detritus in which they live will be ingested by the amphiurans; and since these brittle stars are often lying so densely that their arms can search every square metre of the bottom probably every twenty-four hours, they must be a serious menace for such young stages, which at the end of the larval high season settle on the bottom, in huge numbers.

But this menace of the amphiurans is much reduced when we realise that it is during this high season that they themselves are ready to spawn. When spawning time approaches their gonads swell to four times their normal volume and weight, and yet these gonads must be contained in a body that at other times of the year is occupied almost entirely by the stomach. We know from direct observations that in some species of brittle stars the stomach is constricted and the stomach wall non-functional during this period. For other species it has been proved that they lie apathetically on the bottom in the weeks following spawning and do not respond to any stimuli. Thus it would seem reasonable to say (and this would apply to amphiurans as well) that their appetite is reduced during spawning time and just after and that in all probability

many small bottom stages may survive because of this mechanism.

Let us briefly discuss another unexpected discovery. While most brittle stars have always been considered predatory, a species of sea urchin found on the smooth sea bed (the thin-shelled, short-spined type called the heart urchin, *Echinocardium cordatum*) was formerly considered completely harmless. No matter what time of the year or day they were dissected, one found the long intestine filled with mud or sand (depending on the type of bottom). On average a heart urchin contains three grams of intestinal content at any one time and the whole intestinal content is definitely renewed at least three times per day, probably more often. If we calculate an average of around twenty heart urchins per square metre (which corresponds well with conditions in the slightly deeper sand and mud bottoms in Danish waters) we find that about six kilograms per square metre of material from the topmost surface

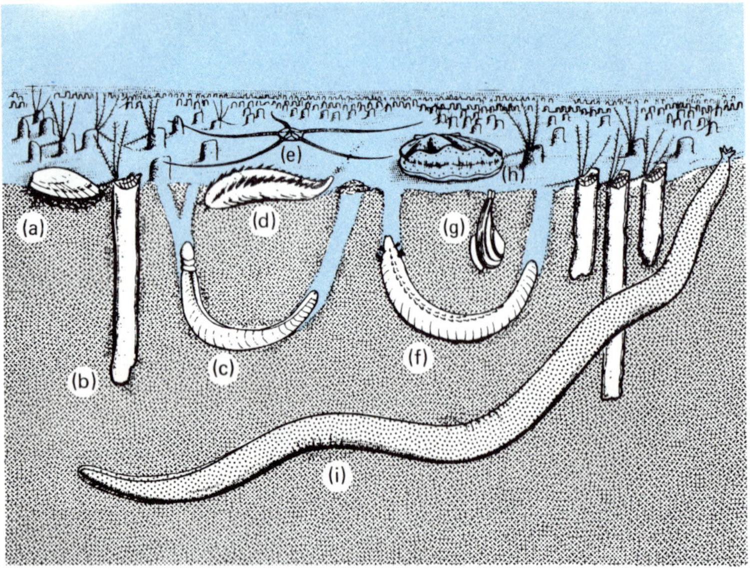

layers of the bottom pass through their intestines each month. This bottom material is unsorted, and is shovelled into the mouth in its natural state. However, as newly settled mussels are no bigger than a grain of sand or a large mud particle, many are accidentally taken into the intestine, killed and digested. An investigation to ascertain how many young mussels are killed in this way revealed that during four weeks in the summer heart urchins actually swallow an average of six to seven hundred bivalves per square metre, and that the majority of them are killed. As compared to the total number of young bivalves settling during this season of the year this is no catastrophe, but the figures do provide a new angle on an animal which had been regarded as a harmless detritus feeder. We have many other so-called harmless feeders, such as bristle worms, holothurians and decapods, which shovel in sand and mud, and in so doing take in a considerable number of young bottom animals in the same indiscriminate manner.

We would, if space permitted, go on to discuss a number of other bottom communities of the sublittoral zone, for instance the *Haploops* community, populated by up to 3,500 small sand hoppers

8·19 Left Vertical section through a Danish clay-bottom *Haploops* community. (**a**) A nest-building bivalve (*Lima*); (**b**) the amphipod *Haploops tubicola* in its vertical leathery tube; (**c**) an enteropneust (*Harrimania*); (**d**) the sea mouse (*Aphrodite*); (**e**) a brittle star (*Ophiura*); (**f**) a bristle worm (*Eumenia*) in its tube; (**g**) a bivalve (*Nuculana*); (**h**) a scallop (*Chlamys*); and (**i**) the hag fish (*Myxine*). (After G. Thorson 1968.)

8·20 Below Mud bottoms in the Skagerrak are inhabited by a *Maldane-Ophiura sarsi* community (bristle worm and brittle star respectively). In one quarter of a square metre we also find heart urchins (*Brissopsis*, an adult in centre of picture), some hundred and fifty bivalves and about a hundred bristle worms of many different species. (After C. G. Joh. Petersen 1918.)

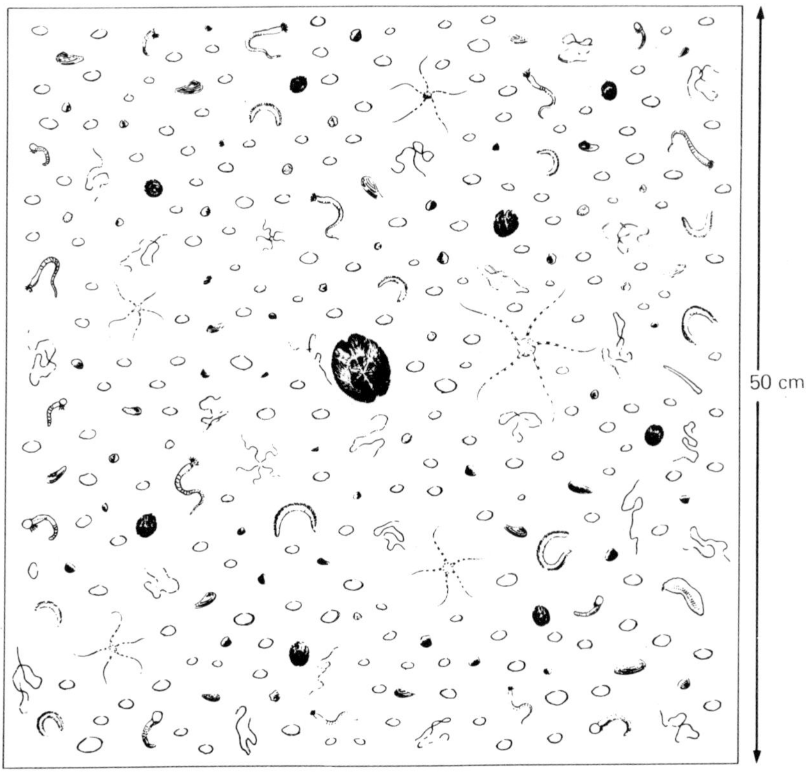

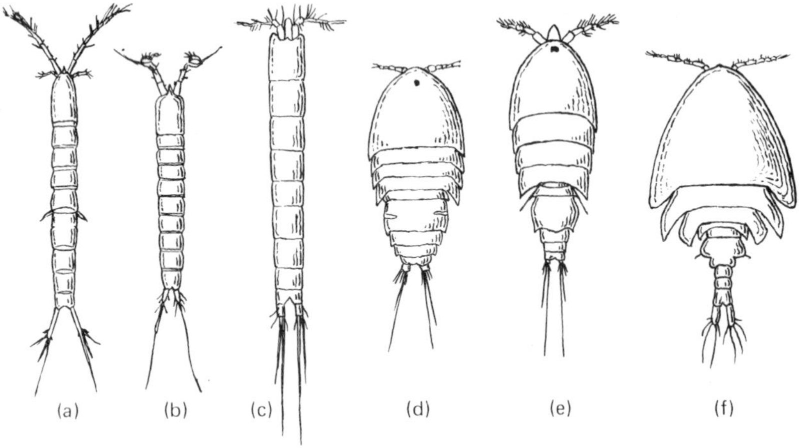

(a) (b) (c) (d) (e) (f)

per square metre, each animal in its own vertical tube; also the *Foraminifera* communities, inhabited by minute, single-celled animals whose pseudopodia stretch for surprising distances and have an amazing catching ability; and in the coarse shell-sand bottom, that most primitive vertebrate, the lancelet (*Amphioxus*), which occurs in such quantities that the stomachs of the local cod populations are often found to be stuffed with them.

Meiofauna

Animals between 0·5 and 2 millimetres in length living in the sublittoral zone are called *meiofauna*. Many are important sources of food for the youngest bottom stages of the *macrofauna*. Looking at a few figures for the meiofauna, free-living round worms, or nematodes (animals of the same group as the human round worm), may number over one million individuals per square metre on shallow water bottoms. Small crawling sand fleas of the harpacticoid type can, living on sand bottoms off the Danish west coast, number over 400,000 individuals per square metre; foraminiferans on the mud bottom in the Sound off Elsinore about one million per square metre; and so on. Many of these small animals not only have many more generations per year than the macrofauna, but also each micro-animal has a comparatively larger surface, which

8·21 Some examples of the meiofauna. The copepods from a sandy bottom (**a–c**) are slender and adapted to life in the spaces between sand grains. Those living on algae (**d–f**) are broad, flat and strongly built, and are adapted to hold on to vegetation. (After G. Thorson 1950.)

means a higher metabolism and a larger food demand. A meiofauna weighing 10 per cent of the macrofauna's weight on the same bottom usually needs a food supply equal to 40 per cent of that of the macrofauna. Nevertheless, it appears that, if we omit the shallowest areas, we can calculate that the total meiofauna production per year does not approach the production of macrofauna in the same area. We must also realise that in many cases half the weight of the meiofauna consists of newly settled young of the macrofauna, which usually grow at such speed that they soon grow out of the meiofauna category and qualify as macrofauna.

On large algae, growths of small algae, polyzoan crusts and polyp colonies house a multitude of small animals, among which a group of crawling copepods, Harpacticidae, occur in large numbers. Ostracods and mites (Halacaridae) are also conspicuous. It would be exciting to view the surface of the micro-algae through the eyes of these small animals; a walk over a polyzoan colony would then appear a dangerous undertaking. Many polyzoa have a kind of defence apparatus called avicularia, or pincers, which seize uninvited guests and pinion them until they die. To pass between the individuals of a colony of hydrozoan polyps often means certain death. Every single polyp is a ferocious predator. Moreover, crusts of polyzoa and polyps such as these can often cover up to half the surface of large algae.

In spite of the dangers of their habitat, the animals of the meiofauna often inhabit sea weed in enormous numbers. A calculation from inner Danish waters shows that a 30 centimetre high plant of the serrated wrack, *Fucus serratus,* can be the home of roughly 25,000 meiofaunal animals – mussels, snails, ostracods, copepods, mites, etc. – during the optimal season in late summer, and to this must be added some 1·5 million individual polyps of polyzoans forming crusts on the surface of the alga.

But if the algae are covered by just one millimetre of detritus or silt, which often happens in estuaries and along the brackish parts of the Swedish west coast, then there will be a different microfauna and meiofauna associated with the sediment bottoms. Such a thin layer of sediment is sufficient to make the plants act as a mud bottom for these tiny creatures.

Microfauna

The microfauna of the sublittoral has so far been best investigated in shallow water. When we sieve quantitative bottom samples through a 2 millimetre mesh, only the larger animals remain: the great majority of very small individuals are lost. Thus the question is, how representative is our microfauna of the total production of the bottom area? What is the role played by these micro-animals,

8·22 A section of the smaller macro- and meiofauna on a brown-alga leaf, *Fucus vesiculosus*. In the background, reading from left to right: a hydroid (*Dynamena*) and (below) a polyzoan colony (*Membranipora*); a skeleton shrimp; a barnacle (with copepods and mites); and another hydroid (*Clava*). In the foreground are several tube worms (*Spirorbis*) and a filtering barnacle (*Balanus*). (After G. Thorson 1950.)

less than half a millimetre long? How numerous are they? How great is their yearly production? How much food do they demand?

Let us start with the smallest organisms – the bacteria. We know that one mud bottom in the West Indies contained around 160 million bacteria per cubic centimetre. But this is merely a 'snapshot'. It has been calculated that marine bacteria can double their number every 1·25 to 8·3 days; these figures applied to bacteria living in the free water masses, and it may well be that those on the sea bottom multiply at a quicker rate. The next links in the food-chain would be bacteria-eating infusoria, which are eaten by predatory infusoria, which are again eaten by superpredatory infusoria. All three types are found on every normal sea bed, and we are thus already up to the fourth link in the food-chain with its attendant loss of nutriment. The small infusoria produce 400 to 800 generations per year and may, in inner Danish shallow waters, occur in numbers from 600,000 to 20 million specimens per square metre. Diatoms have 150 to 200 generations per year and in the same waters amount to 4 to 11 millions specimens per square centimetre. As far as accurate calculations are concerned, we are, however, still on dubious ground.

9 Below two hundred metres

The bathyal zone

We have so far referred to the area from low water mark to about 200 metres as the sublittoral zone; it is also known as the continental shelf – a term for the flat land areas which continue under the surface of the sea. In most cases the shelf slopes gently, but at about two hundred metres depth the inclination suddenly becomes steeper and the sea bed falls sharply towards the deep sea proper. This steep zone is called the *continental slope;* it is also known as the bathyal or archibenthal zone, which at a depth of 2,000–3,000 metres passes into the smoother deep sea bed. The continental slope itself is usually far from smooth. It can be furrowed by immense fissures, the so-called submarine canyons, and the epifauna and infauna are seldom clearly separated.

Excellent places for studying the fauna of this area are the deep Norwegian fjords. Sogne Fjord, for example, has depths down to 1,200 metres and here, close to the land, one is able to reach the many strange animals which have chosen this zone as a habitat. It was off the coast of northern Norway in 1864 that the famous zoologists M. and G.O. Sars found for the first time the small stalked sea lily, *Rhizocrinus lofotensis*. Hitherto, this ancient group of echinoderms had been known only as fossils from geological deposits but now there was proof that stalked sea lilies still lived in the depths. The discovery caused great interest among zoologists and helped to promote the first deep-sea expeditions. Another creature (often with a two metre long stalk) from these regions is the luminiferous eight-armed coral *Umbellula*. All the polyps are gathered into a cluster at the top of the stalk, so that the animals closely resemble a large lily-like flower with a bowed head. Another eight-armed coral, *Paragorgia,* coloured a brilliant orange-red, is one more of the attractions found lower down the slope. In the Norwegian fjords the 15 centimetre long mussel *Lima excavata* is

9·1 Vertical section through a deep-water community in the Skagerrak at two hundred metres. (**a**) The red shrimp *Pandalus borealis*; (**b**) the brittle star *Ophiura sarsi*; (**c**) the grenadier fish (*Coryphaenoides*) with two sea pen colonies (*Funiculina*) in the background; (**d**) hunting shrimps; the bristle worms (**e**) *Myriochele* (thread-like), (**f**) *Melinna* (tentacles in front), and (**i**) *Maldane sarsi*; and four bivalves (**g**) *Syndosmya*, (**h**) *Nucula*, (**j**) *Thyasira*, and (**k**) *Nuculana*. (After G. Thorson 1968.)

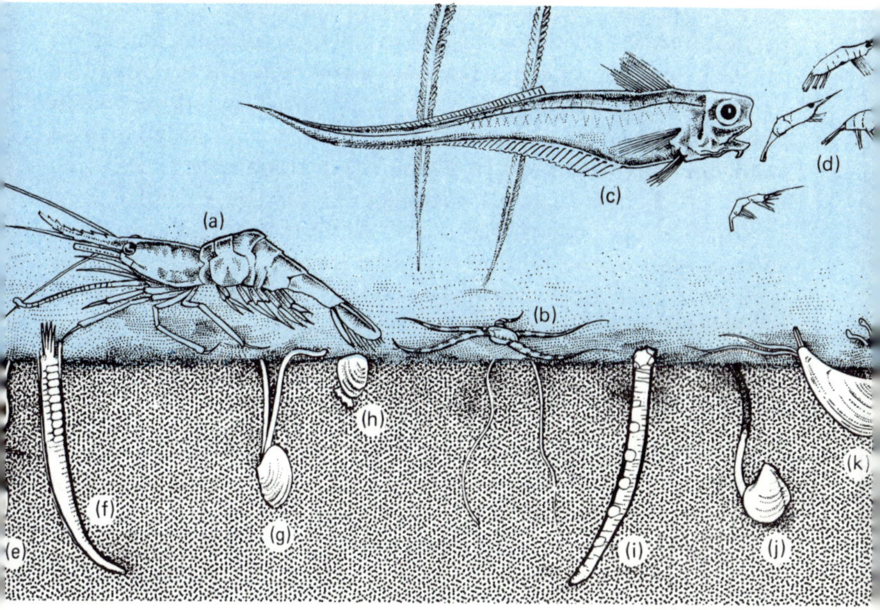

characteristic of this society. It swims by flapping its shells, and builds around them a nest of horn-like, silky threads, which it secretes itself. It is also here that the first siliceous sponges occur. We can mention, in particular, the glass sponge, *Euplectella,* which resembles a 30 centimetre long slender vase of plaited glass anchored into the sea bed with a root network of siliceous needles – a beautiful sight indeed. Other siliceous sponges can be almost one metre high.

Among the fish of the continental slope one should mention the rabbit fish, *Chimaera.* This creature has a large head, very large eyes, almost wing-like pectoral fins, and a whip-shaped tail; it lives on

9·2 Sea cucumbers (Elasipoda, genus *Scatoplanes*), reminiscent of grazing cattle, and brittle stars (*Ophiomusium*) characterise the mud bottoms off southern California at 1200–1300 metres depth.

bottom animals and is closely related to sharks and rays. A family of whip-tailed fish, Macrouridae, is common in this zone; despite their appearance, the family relationship to *Chimaera* is very distant, and they are in fact closely related to cod fish. All in all, animal life on the slope is surprisingly rich and varied, and it is here that one finds most representatives, geologically speaking, of the oldest animal groups. But as an example of the decrease in number of individuals with increasing depth one can mention some figures from the latest French deep-sea investigations in the Atlantic, using trawls. Between 200 and 500 metres depth the average number of fish per haul was around two hundred individuals. Hauls between 4,000 and 5,000 metres yielded an average of only twenty-one fish per trawl.

The abyssal zone

This zone comprises by far the largest area of the deep sea bed proper, from 2,000 to about 6,000 metres depth; 84 per cent of the ocean bottom (more than half the surface of our globe) is deeper than 2,000 metres and 82 per cent is at depths between 2,000 to 6,000 metres. Here we are down in a very stable milieu, where the temperature never exceeds 4°C. About one-half of the abyssal area is covered by the calcareous shells of globigerina and coccolithophores; about one-seventh is covered by skeletons of diatoms and radiolarians; and the remainder, below 4,000 metres, is covered by fine red deep-sea mud, with a very small content of organic material and a scanty fauna.

The number of animal species, many of them cosmopolitan, is significantly reduced and is separated into two sharply divided groups – mud eaters and predators. In the first group are the holothurians, which become more and more dominant the deeper one goes. These small echinoderms, especially of the group *Elasipoda*,

are extremely thin skinned at this depth, often almost glass clear; to a large extent they are composed of water, their content of organic dry matter being extremely small compared with their live weight. Many have long appendages along the sides of the body, with the help of which they crawl slowly over the soft bottom, simultaneously shovelling large amounts of material into their mouths. Their small dry-matter content is an indication that their metabolism is very low; and as it is suspected that they grow very slowly and live long, they should have little difficulty in maintaing themselves in this nutriment-poor milieu. A genus of mud-eating brittle stars, *Amphophiura,* can also be numerous on the sea bed, and even among the sea stars (Asteroidea), usually regarded as pronounced predators, we have groups like the family Porcellanasteridae which appear to concentrate on mud. The same is true of a number of abyssal polychaetes of which, however, many are almost cosmopolitan and can also be found at much lesser depths. Sea spiders are also found here; the species *Colossendeis colossea* in particular is impressive in having a leg span which can measure over half a metre. We know very little about its food, though we do know that there are caeca (lengths of blind gut) stretching from the rather narrow middle intestine far out into the legs. The predators proper of these depths include some sea stars, a number of fast-moving crustaceans and, above all, fish of the families *Stenoptyx*, *Gigantura* and *Stylephorus*.

The question of balance between predators and prey in the abyssal zone is one of the unsolved problems of the deep sea. Figures from the latest Russian expeditions indicate 2 to 5 grams live weight of bottom animals per square metre at 4,000–5,000 metres depth, but samples from the *Galathea* Expedition were as low as 1 gram per square metre, and from some areas we have still smaller weights. As mentioned, the predominating animals, holothurians, have a very small content of dry matter – hardly above 2 or 3 per cent of their live weight. The question is, when we now find 1 to 5 grams per square metre, what does this mean calculated in terms of dry matter? How many years has it taken for these animals to become adult? How quickly can the fauna renew itself? What is the yearly production? If we calculate the dry matter weight to be 2–3 per cent of the live weight, we then have only 20 to 150 milligrams dry matter per square metre and if this amount has taken ten to twenty years or perhaps even fifty years to accumulate, we are dealing with very small figures.

It would appear that bacterial production on the deep sea bed lies in a very different order of magnitude. Down in the Philippine Trench the American bacteriologist Claude Zobell found a production of 8 or 9 grams of organic carbon per year per square metre, and even though errors of calculation can be great, we still have values that are completely different from those shown by the macrofauna.

However, there can be three links of infusoria between bacteria and multicellular animals in the food-chain, which means that bacteria may be utilised as food for macro-animals only to a small degree; this makes the contrast more understandable. But we have other ways of evaluating the production rate on the bottom. One of these is the predator test, which investigates how far animals which must find living prey penetrate into the deep sea. We find that crabs have never been caught below 5,200 metres; brittle stars stop at 8,006 metres (the deepest living species being probably mud eaters); sea stars at 7,630 metres; and fish at 7,580 metres; though many other groups which are omnivorous or are at least not dependent on living prey go right down to 8,210–10,700 metres.

When a predator hunts there must be a certain relation between the energy expended in hunting and the size of the booty. If prey is too widely distributed the hunt results in an energy deficit and the hunter will eventually perish. It would appear reasonable to assume that the larger predators simply do not find enough prey in the deepest parts of the sea. The figures we quoted above: 1–5 grams live weight per square metre, were from 4,000–5,000 metres depth. In samples deeper than 2,000 metres in the north-west Pacific the Russians found only 50–100 milligrams live weight per square metre of bottom, and in corresponding samples from depths of 9,000 to 10,000 metres only 20 to 30 milligrams. The minimum weight that would permit the existence of larger predators appears to lie somewhere between these weights.

It seems, however, that all weights calculated so far for the bottom fauna of the abyssal have been falling short of the truth. New, careful collecting and sorting methods have yielded many more species, mainly of isopods, amphipods and bristle worms; these were probably overlooked because of their minute size and fragile structure. The American biologists R. R. Hessler and H. L. Sanders, in a haul from 4,700 metres depth in the Sargasso Sea, found no less than 196 species of animals, and many individuals of some of them. The meiofauna too seems to be somewhat richer than might have been expected. Thus H. Thiel, during a cruise of the *Meteor* in the Indian Ocean, found 16 to 170 meiofaunal specimens per ten square centimetres at depths between 1,000 and 5,800 metres, which is more than has been found in shallow brackish waters around southern Finland. These new findings, although of the greatest interest, will, however, hardly change the general impression of a very low weight and a small productivity in the deep sea, since most of the species are tiny and probably grow very slowly.

During the first decades of deep sea investigations it was agreed that the stable conditions in the abyssal must provide favourable habitats for the – geologically speaking – older animal species. It was generally believed that if 'living fossils' existed in the sea, it was here that they would be found. This view is no longer held; it seems that on the whole they are more likely to be found in the bathyal zone, between 200 and 1,000 metres. When the Coelacanth *Latimeria* suddenly made an appearance off the south-east coast of Africa in 1938 (to the amazement of those who believed the whole group had died out 200 million years ago), it was at first thought that it was a deep-sea fish; but subsequent catches showed that they actually live in comparatively shallow waters. In fact it seems that the whole abyssal fauna is relatively recent and that it originates from forms which have migrated down from the bathyal. It is probable that the abyssal fauna has had so short an existence in geological terms that it has not yet reached full development or established its identity from a geological point of view.

Nevertheless, it was found during the *Galathea* Expedition that the abyssal zone can produce surprises. In one trawl haul of 3,590 metres off the Pacific coast of Mexico there appeared some molluscs with flat, limpet-like shells which, at first glance, did not appear to be very exciting but were nevertheless preserved for later examination.

It was not until well after the return of the expedition, during the detailed sorting of the material, that Henning Lemche revealed them as the greatest sensation of the whole expedition, the primitive mollusc *Neopilina*, treated monographically in the *Galathea* reports. This shield-shaped mollusc with its primitive structure hit world headlines, and as a zoological find it fully ranks with the Coelacanth. In recent years, other species and new genera of the group *Monoplacophora*, to which *Neopilina* belongs, have been described, and future deep-sea expeditions will probably add further species to this fascinating group.

The hadal zone

The ultra-abyssal zone, which comprises the area from 6,000 metres to the bottom of the deepest trench, but which covers less than 2 per cent of the ocean bottom, is called the hadal zone. It is certainly populated by living creatures, although these are very poor in numbers as compared even with the abyssal. It must, however, be remembered that our knowledge of this part of the ocean, from which the first trawl haul was taken as recently as 1948, is still very limited.

The late Anton Bruun suggested, on the basis of his own material, and of material from Russian expeditions, that the hadal should be regarded as a separate zone with its own special animal life. The hadal zone is restricted to the deep trenches, most of which were studied only quite recently, starting with the Swedish deep-sea expedition, followed by *Galathea,* and in recent years on a large scale by the Russian research vessel *Vityaz*. At present we know of twenty-two trenches with depths exceeding 7,000 metres. Six of these – Marianas, Tonga, Kurile-Kamtchatka, Japan, Philippine and Kermadec – exceed 10,000 metres. Furthermore, we know around twenty more trenches with depths between 6,000 and 7,000 metres. The water temperature of all these trenches is from 1·2°C to 3·6°C – somewhat lower than those in most parts of the abyssal. The trenches that lie close to land (and most of them do) have a richer animal life than those in the open ocean. A calculation from 1966 gives the total number of species taken from depths exceeding 6,000 metres as at least 366. Nevertheless, this is a very small number as compared with those from the abyssal and the continental slope. The Swedish deep-sea expedition found in their trawl hauls from 2,500 to 4,000 metres an average of fifty to sixty species, and from 8,210 to 8,300 metres an average of twenty species per haul. The Russian expeditions in the north-west Pacific calculated that trawl hauls from 1,000 metres brought in an average of 120 species of bottom animals, while three quite similar hauls at 8,000 to 9,000 metres contained only six, eight and seventeen species respectively. Here, most of the larger predators, from fish to crabs and sea stars, are completely absent, and the deeper we go the more dominant become the mud-eating sea cucumbers (holothurians) which constitute about 90 per cent of all animal weight in the deepest samples. A single trawl haul in the Sunda Trench at around 7,000 metres contained 3,000 holothurians. Sea anemones are also encountered at the greatest depths. These, too, have a very low content of dry organic matter, and belong to a group in which individuals have been known to live fifty to sixty years. In addition, there are some sixty to eighty species of bristle worms, well over a hundred species of crustaceans, some quite small mussels and a larger number of representatives of a strange group of thread-like worms, *Pogonophora,* which lack among other things, a gut. Here again it is assumed that the fauna is composed of comparatively new immigrants from the abyssal layers which have still not reached full development. But the deep trenches lying as isolated troughs may also have an endemic fauna, which can live there and nowhere else. Life, however, does exist, despite the distance from the surface being more than

10 kilometres, the enormous pressure, the poor food conditions and the total darkness. Perhaps the very fact that the larger predators cannot live here is the attraction – it gives the least demanding animals a peaceful habitat in the hadal zone.

10 The sea as a source of food

We conclude this short description of life in the sea by discussing its relevance to man. At the beginning of this book we said that the sea gives us less than 2 per cent of our food. While the world's catch of fish, crustaceans, molluscs and other invertebrates has more than doubled in the last thirty years, the all too familiar population explosion has more or less nullified the increase and the percentage of the sea's contribution remains deplorably low – or so it would seem. The question of food from marine sources is made doubly important by the fact that the sea has so far provided just that kind of food which is so lacking in the diet of the world as a whole, namely protein; and there is a crying need for more effective exploitation of a habitat which is, as we said, three hundred times greater than that of the land and fresh water put together. Surely there must be some way in which this vast habitat can be put to better use. But, as we shall see, there are complications.

Primary production

Let us start at the bottom of the food-chain, with the primary production of phytoplankton. Twenty years ago, a marine botanist would have speculated that the plant production of the seas was roughly ten times that of the land. Today we are less optimistic, largely due to the pioneer work of the Danish phytoplankton expert E. Steemann Nielsen, who devised a new method of measuring phytoplankton by assessing the quantity of carbohydrates produced by micro-algae when exposed to carbon dioxide in which the carbon atoms consisted of the radioactive isotope carbon-14.

This new method was extensively used during the round-the-world expedition of the *Galathea,* and the results were somewhat depressing. It was shown that, far from being ten times as productive as the land, the sea produced approximately the same quantity of plant matter as the land, yielding some 15,000–20,000 million tons of organic carbon per annum. But since the area of the sea is at least twice that of the land, this of course means that the sea's productivity per unit area is about half that of the land.

In comparing sea and land productivities, we must remember that while experts calculate that the land can be made to at least double its production within the next fifty years by means of irrigation, fertilisers and improved breeding of crops, there is absolutely no prospect of being able to step up productivity of the open oceans. The most we can hope for would be some improvements in 'marine agriculture', in shallow water coves and land-locked fjords.

The sea thus accounts for about 50 per cent of the world's primary production, and for this reason alone it ought to be more effectively exploited than it is at present. But there is a huge discrepancy between the sea's primary production and the harvest of animals that man takes from the sea. In 1966 the catch of all marine animals was 49 million tons. Comparing this with the calculated primary production of 15,000–20,000 million tons, we are faced with the realisation that there is a 400-fold difference between primary production and catch. But if calculated as millions of tons of carbon, we can say that the difference between catch and primary production is really 4,000-fold.

Even where there is a highly productive fish crop such as the huge assemblage of anchovies off the coast of Peru, where the annual production is known to be eighteen million tons, the annual catch is only nine million tons. Anchovies live directly on phytoplankton, so that here the food-chain is as short as it can be.

Food potential

How large a yield can we hope to gain from the oceans in the future? Even the most expert calculations are suspect, but the best estimate that we have up to 1969 is that we should be able to take 200 million tons – four times the present catch, without depleting the population. In terms of the human population explosion this estimate is not very encouraging; nevertheless, as we have said, the protein from sea food is extremely important in countries with a low standard of living.

While regretting this relatively small harvest from the sea, we must remember that there is no possibility of 'cultivating' the sea on a large scale; even if there were, fishermen would hesitate to fertilise an area of open water from which they themselves would in all probability not reap the benefit. Fishermen are keen to improve their boats and methods of fishing but they cannot avoid remaining at the Palaeolithic level of being hunters and not farmers. The result is that our utilisation of the products of the sea lags about two hundred years behind our achievements with land resources. And to make matters worse, man as a hunter is not particularly successful compared with marine animals. He cannot, like the copepod or the euphausiid, catch phytoplankton efficiently; nor can he catch copepods profitably, like the herring (to mention only one fish). Much has been made in recent years of the statistics of squids caught for human consumption (approximately 600,000 tons) but no one has yet succeeded in netting the giant oceanic squids which are numerous enough to support a considerable population of sperm whales. And on the subject of whales, the smaller toothed whales regularly catch much larger specimens of horse mackerel than Japanese fishermen (themselves experts in this kind of fishing) ever succeed in taking.

Too much time is spent in hunting for fish and too little in catching them. Fish populations change their locations from season to season and even from day to day. It is true that when a good fishing place has been found the boats are quickly filled, but then they must return to their home port, when they could have secured a larger catch if only they had had better storage facilities.

Farming invertebrates

If we consider those invertebrates that are acceptable for human food, we find that these operate far down the food-chain. Let us therefore turn our attention to the possibilities of farming shrimps, prawns and mussels.

These invertebrates may well turn out to be one of the greatest contributors of protein. At the present time fish constitute about 89 per cent of the world catch of marine food, while the lower animals we have just mentioned amount to only about 11 per cent. Clearly there are great possibilities of increasing the latter figure, particularly as the lower animals flourish in shallow water, utilise their food better, grow faster and are easier to control and to harvest.

The full development of invertebrate farming presents a challenge to marine biologists, if optimum conditions are to be established. We have an interesting example of what can be done in some recent work that has been carried out by Russian biologists in the Caspian Sea. This sea was, of course, part of the ocean millions of years ago, and although now isolated it still has a marine fauna. On investigation, however, the Caspian was found to contain a poorer fauna than comparable areas, such as the Black Sea and the Sea of Azov. In particular the bristle worm *Nereis diversicolor*, and the mussel *Syndosmya ovata*, both excellent fish food, were completely lacking in the Caspian, though plentiful in the Sea of Azov.

Both these species were introduced into the Caspian, with such effect that the most populated areas are carrying seven thousand *Syndosmya* per square metre, while *Nereis* is even more prolific. The net result is that the production of bottom animals has increased many times and has greatly stimulated fishing in the northern part of the Caspian.

Shrimps and prawns No account of marine food resources would be complete without special mention of the very extensive inshore fisheries of shrimps and prawns in the United States. These crustaceans are highly nutritious and form a large part of America's protein harvest from the sea. For example, the 1965 crop of the Atlantic and Gulf of Mexico fisheries was valued at 82 million dollars and exceeded the Pacific salmon fisheries by 22 per cent. Japan, Thailand, Germany, Canada, Spain and France also produce a high yield of shrimps, while Denmark and Norway are slowly beginning to develop deep-water shrimp fishing in the Skagerrak, off Spitzbergen and in the waters off west Greenland. In 1937 the Danish fishery biologist H. Blegvad discovered the existence of a large shrimp population in the Persian Gulf, thus paving the way for a new fishing industry.

The reasoning behind the farming of fish under controlled conditions can be applied to shrimp culture. In South Carolina shrimps have been induced to yield fourteen to fifteenfold in two months. Similarly in Galveston Bay, Texas, experimental yields of twenty-seven to twenty-eight tonnes per square kilometre have been quoted and there are culture ponds in India where the yield has varied from 30 to 120 tonnes in a season of only five months. Tropical prawns, especially those of the genus *Penaeus,* often reach impressive sizes. The banana prawns of north-western Australia, for instance, can grow to a length of 27 centimetres and lengths of about 18 centimetres are quite common. A Japanese species of *Penaeus* can be kept in mass cultures in laboratory tanks until they are four to five centimetres long, and then transferred to large ponds at about 10°C. Here they mature in one year with a length of about 12 centimetres and a weight of about 25 grams.

Cultivating and fishing for bivalves Surprising as it may seem, bivalves already comprise two-thirds of all invertebrates taken in organised fishing. In many areas the bivalve harvest greatly exceeds that of fish. For instance, the 1952 statistics for North Carolina gave a figure of about 7 million tonnes for fish, but more than 15 million tonnes of shell fish. The catch of sea scallops (*Pecten*) in the United States amounted in 1961 to 12,480 tonnes, measured as scallop meat, that is without shells, while in 1964 the yield from the Atlantic coast of the meat of the surf clam (*Spisula solidissima*) amounted to 17,000 tonnes (again measured as meat). The surf clam fishery along the New York–New Jersey coast in 1961 was valued at 20 million dollars. This is impressive, but the yield is being still further improved by the use of more sophisticated gear. Large clams have often escaped capture because they were buried too deeply in the sandy bottom to be caught by the old-fashioned dredge. Now, however, the dredge is fitted in front with a row of iron pipes, like the teeth of a rake, and these pipes discharge strong jets of water, which disperse the sand and expose the clams to the approaching dredge.

Certain countries, such as Japan, the United States, Holland, Spain and France, in that order, exploit mussels. Denmark is active in both oyster and mussel fishing, particularly of the common mussel, but the Norwegian and Swedish mussel catch is too small to appear in international statistics. This is regrettable, because biologists are agreed that no animals can produce as much food for human consumption in as short a time as bivalves in general, and that the common mussel (*Mytilus*) is probably the most productive of all. The concentration of mussels on tidal flats is very great; in Denmark for example the weight of living mussels can be as much as 36 kilograms per square metre. Another computation indicates that the flats of the southern Baltic carry over four million tonnes of mussel meat (excluding shells). The mussel, like the copepod, is a 'key industry' animal that converts plant material into high-grade protein. Furthermore, its powers of propagation are colossal; a medium-sized common mussel (*Mytilus edulis*) produces seven to twelve million eggs per annum. This egg-laying capacity is, however, exceeded by the American oyster's 100–500 million.

So far we have dealt with bivalves in their natural habitat, but here again the prospects of farming them seem to be even more promising than the prospects for crustaceans. Here we are indebted to the American oyster expert, V. L. Loosanoff, who has elaborated a technique for conditioning and stimulating clams, mussels, and oysters to spawn in tanks independently of the seasons. He has also managed to rear mass cultures of larvae and of the newly settled spat. Baby clams from New England hatcheries can now be transported in large quantities to be laid out, for instance, on banks in Florida where, in the warm water, they can grow to marketable size much more quickly than in their native habitat. Similar results may be obtained in England, by placing hatcheries and farms in estuaries close to the cooling water outlets of large power plants, especially nuclear power stations, which use sea water for cooling. Here an artificially warmer climate can be established, which allows a quicker cultivation and growth of bivalves, shrimps, and fish. Incidentally, adult oysters are now known to grow fat and taste delicious on a diet of cornmeal. Thus the prospects for bivalve culture seem to be very favourable.

Of all 'fishermen', oyster and mussel farmers employ methods that most closely resemble those of land agriculture. Very large coastal areas are systematically cultivated and tended and we have proof that well-managed oyster banks can yield up to ten times more than comparable areas that are not cultivated. But it must be understood that it is not enough to merely cultivate and take care of the mussels; it is equally important to know the biology of their enemies and competitors for food as well as knowing their diseases. Enormous sums and much ingenuity have been employed to combat the threat of sea stars and boring snails on mussel banks and new methods are constantly being tried. In some areas noxious animals are ploughed

10·1 A possible scheme for the utilisation of solar energy, flue gas and cooling water effluents from industrial sites (for instance, atomic plants) for algal and shell-fish culture. (Redrawn from A. D. Ansell.)

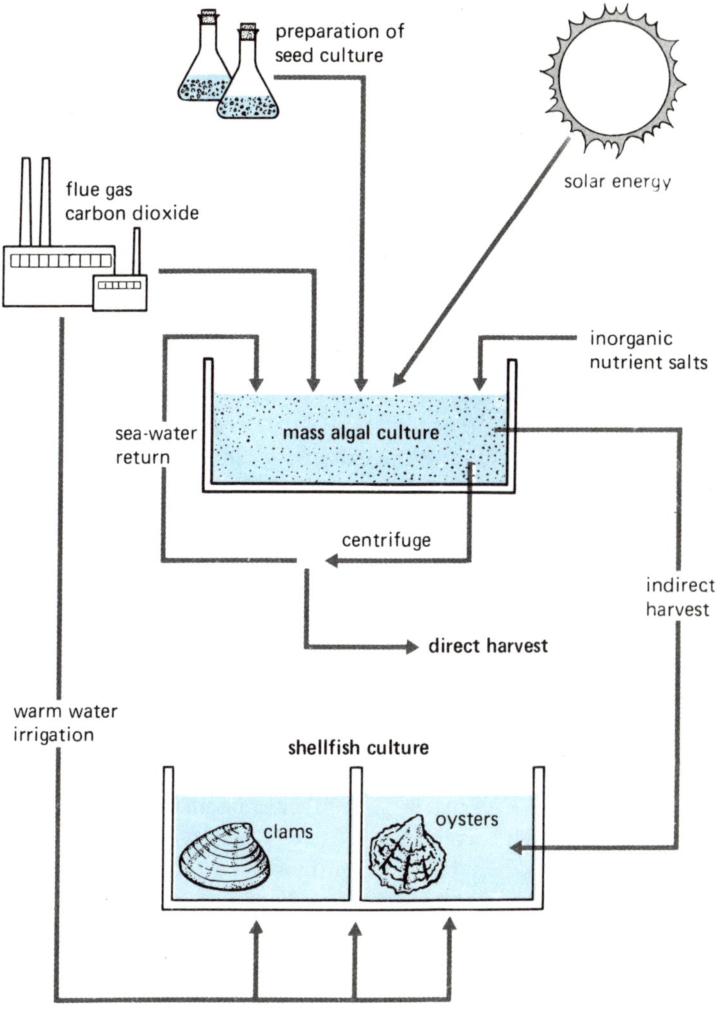

so deeply down into the sea bed at low tide that they can never reach the surface again, thus ensuring that the bottom has been cleared sufficiently to receive new bivalve broods. In other places the bottom, and thus all the predators living on it, is covered by kilometre-long layers of plastic sheet on to which bivalve larvae have no apparent reluctance to fasten themselves. Another method is to shovel slaked lime over the sea stars; these then disintegrate while the bivalves, protected by their shells, come to no harm – an expensive but effective method. Finally, the sea bed has also been directly 'vacuum cleaned' to eliminate sea stars. All these methods are fairly successful, but in the sea we must above all exploit nature's own means; and in order to do this one must know one's animals. To illustrate this, here is a striking if somewhat complicated example. The sea star has a period of about four weeks just prior to spawning when its appetite is lost or at least slowed down. The caeca of its stomach, which lie in the thick arms, are pressed together by the swelling testes and ovaries, and are more or less unable to receive food. Thus if the bivalves begin their life on the bottom at the same time as the sea star loses its appetite, they have a whole month in which to grow before being hunted. One month after attaching itself to the sea bed a newly precipitated bivalve will have increased its volume about five hundred times. This means that when the sea star regains its appetite it will obtain just as many calories by eating one bivalve of this size as it would have gained by eating five hundred bivalves a month earlier; in other words the non-eating 'passive' period of the sea star spares the lives of a large number of bivalves. This important breathing space depends on the bivalves reaching the bottom at the correct moment, when they can 'get their full month', and in natural coastal waters this rhythm has long been established by natural selection.

A mussel or oyster farmer often considers himself smarter than nature; he has heard of another oyster type that grows more quickly, is plumper, has a better sales appearance, or has fewer parasites than those in his own coastal areas. He then tries to import such a type, usually without knowing whether it will spawn at the same time as the indigenous population. Often it does not – and the result is that the brood arrive at the bottom right in the middle of the sea star's most voracious period, and the crop is ruined. This is not to say that new strains should not be introduced, but it is important to cross bivalves from different temperature areas in order to transplant a cross-breed that will spawn at the right time.

Fishing

The need for new methods of fishing There is an urgent need to devise new and improved methods of fishing, and not to depend on conventional methods. A former American chief of fishery investigations, Lionel Walford, has stated that the ideal constructor of new techniques 'should have the initiative and understanding of a scientist, draw like an architect, build like an engineer, speak a language that fishermen can understand, and think in the same way as a fish.'

On a more scientific level we need to learn far more about the biology of these valuable sources of protein. So far we have pursued an extremely short-sighted policy, and in the advanced countries there has been too much emphasis laid on research dedicated to practical results. By this we mean investigations into fish which directly and on a large scale serve as human food. The so-called 'unusable' fish have been relatively neglected. And yet it may well be that these unusable animals live on the same bottom and eat the same food as the edible fish, in which case they are competitors. Even more ominous is the possibility that the unusable creatures live on the edible fish and are therefore direct competitors with ourselves. Surely it follows that it is just as important to study their

10·2 The feeding habits of the chief commercial fish of the Barents Sea. Reading from left to right are the haddock, cat fish, sand dab, ray, cod, and sea bass. (After L. A. Zenkewitsch 1963.)

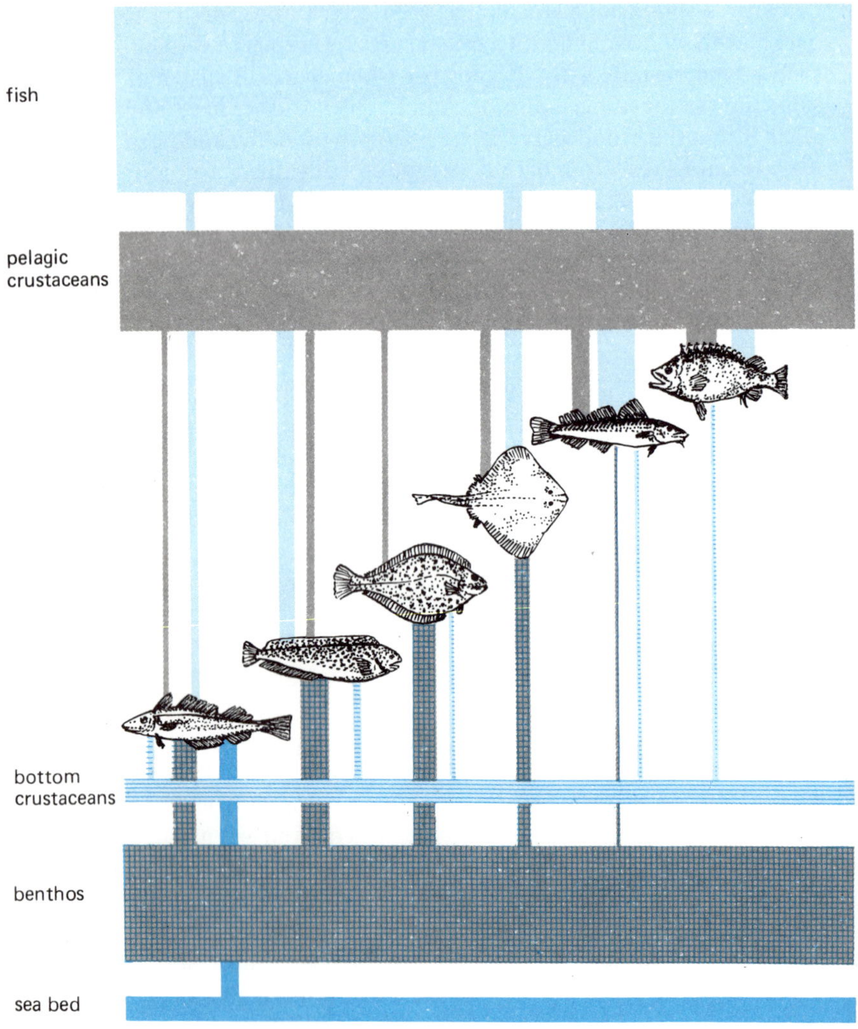

biology so that we may control them as it is to study and develop the production of edible fish.

Shallow water fisheries Open sea fisheries are not really amenable to human control; and we can add that massive water movements and temperature variations make it all but impossible to predict where the best fishing is to be found, and even what any one season's catch is likely to be. However, we do have some say in the development of improved fisheries in *shallow* water and in fjords and coves. One can, for instance, fertilise a fjord that has a narrow entrance, thereby increasing the production of micro-algae and detritus. Such an increase in primary production will automatically be reflected through the food-chain. Nevertheless in so doing we have to ask ourselves whether we are favouring the animals that we want to develop for human consumption. Some kind of an answer can be obtained only by conducting a census of the sea bed population and then by considering how the available food (primary production) is allocated among different animals. Such a census was initiated for the whole area of the Kattegat by Petersen in 1910; subsequent revisions have refined the data to a point where we can claim to have a realistic picture of the sea-bed population at any particular time.

Among the bottom animals that convert plant food into protein, five million tonnes are of no great importance as food for other animals; but another one million tonnes – small snails, mussels, worms and crustaceans – constitute 'fish food' and provide nourishment for all bottom-living predators. These predators can be divided into two groups: bottom fish, especially flat fish (mostly plaice in the Kattegat) amounting to 50,000 tonnes; crabs, predatory snails and sea stars (75,000 tonnes); and brittle stars, with an impressive weight of 300,000 tonnes. Admittedly brittle stars are omnivorous, but it would be a modest estimate if we were to say that one-quarter of their intake was animal prey; and that being so we can regard them

10·3 The standing crop of bottom-feeding fish and bottom invertebrates in the Kattegat (Denmark), for the year 1910. (Redrawn and much adjusted, after C. J. Joh. Petersen 1918.)

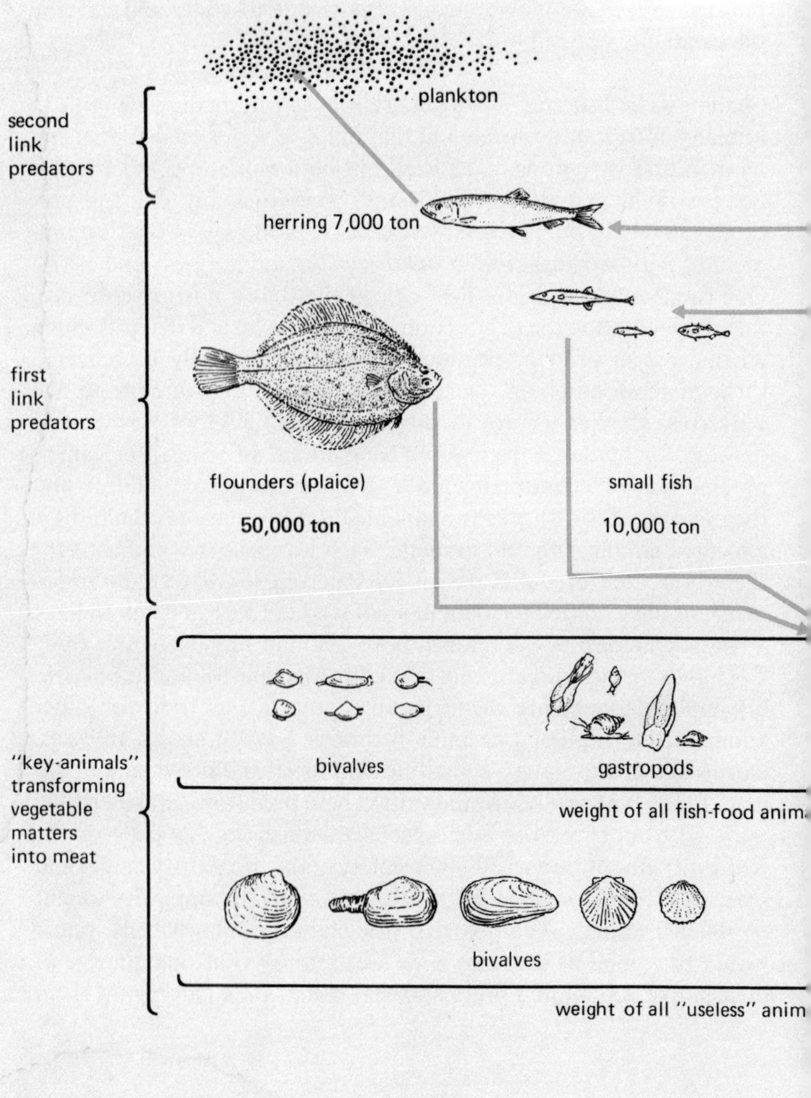

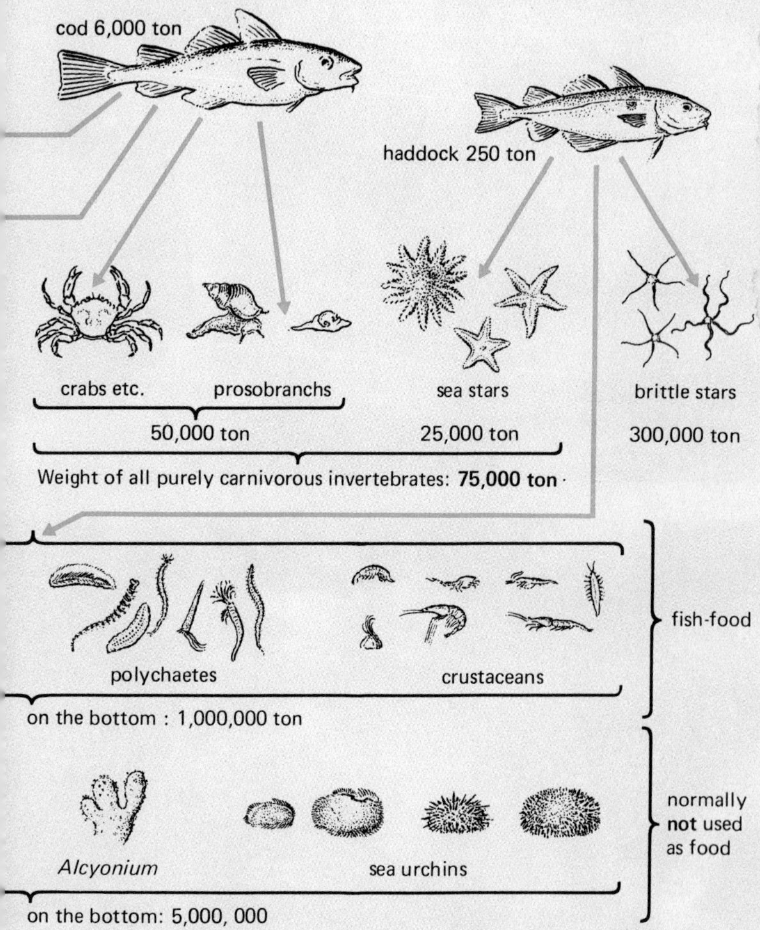
cod 6,000 ton
haddock 250 ton
crabs etc.
prosobranchs
sea stars
brittle stars
50,000 ton
25,000 ton
300,000 ton
Weight of all purely carnivorous invertebrates: 75,000 ton
polychaetes
crustaceans
fish-food
on the bottom : 1,000,000 ton
Alcyonium
sea urchins
normally
not used
as food
on the bottom: 5,000, 000

10·4 An adult brittle star, *Ophiura texturata*, with the dorsal side opened (bottom) to show the size and quantity of the bivalves, *Spisula*, which a few minutes earlier the brittle star had engulfed via its modest mouth opening (top). (Howard Feder 1965.)

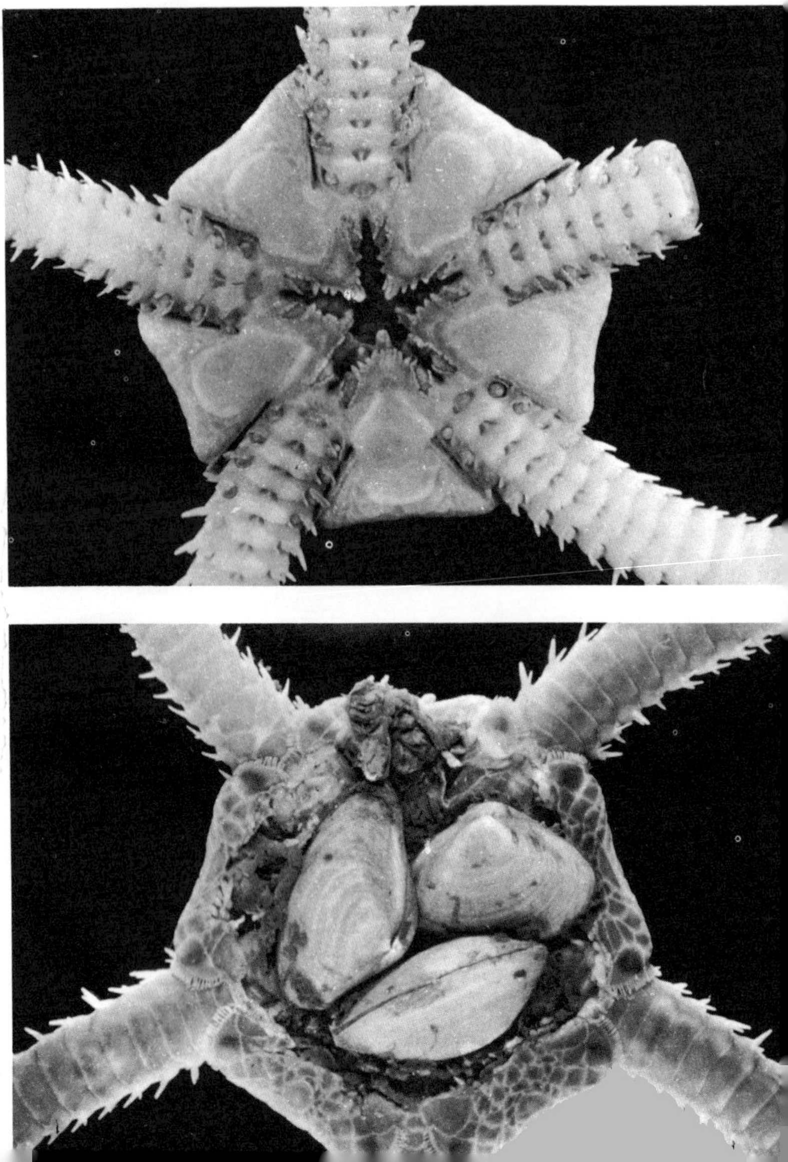

10·5 Top The cat fish (*Anarhichas lupus*) is able to crush the shells of even the largest bivalves which have dug themselves into the bottom.
Bottom The swimming crab, *Portunus depurator*, with its hindmost pair of legs built like oar blades, is one of the predators which feeds on the same invertebrates as the bottom-feeding fish. (After G. Thorson 1968.)

10·6 The volume and weight of the Danish sediment-bottom invertebrates needed to produce one gram of dry organic matter. (After G. Thorson 1957.)

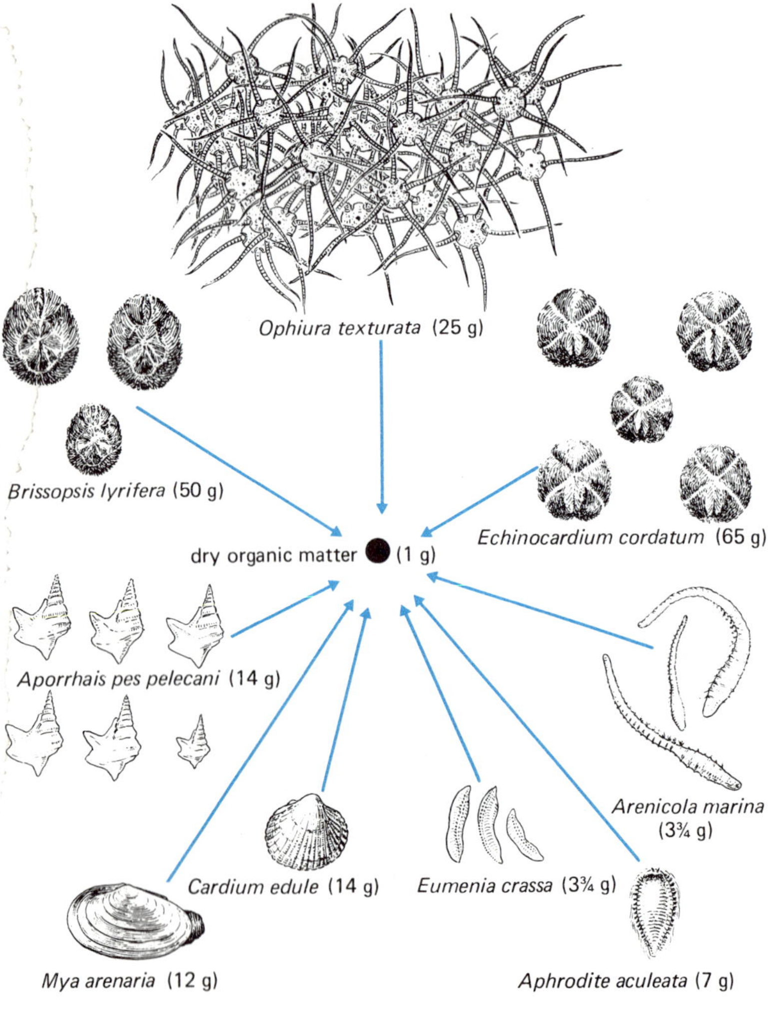

as being equivalent to 75,000 tonnes of predators. The total weight of invertebrate predators in the Kattegat thus works out at about 150,000 tonnes.

Both the fishes and the invertebrate predators share the available one million tonnes. From the above figures we can calculate that if flat fish and predators were equally voracious the available fish food would be divided between them in the proportion of 1 to 3; in other words the flat fish would eat 25 per cent while the crabs, predatory snails and brittle stars would eat the remaining 75 per cent. The next question is whether flat fish and lower predators eat proportionally the same amount.

When Petersen compiled his original census he had no information on this important question, but today we are better informed. We now know that during their most intensive period of growth, such animals as crabs, predatory snails and sea stars, to mention only a few, eat an average of 20 per cent of their own live weight every twenty-four hours. In contrast, a plaice is satisfied with a daily intake of 5 per cent. Such a discrepancy is intelligible when we realise that a plaice takes four to six years to reach sexual maturity while many of the lower predators achieve full growth in one or at most two years; in fact, lower predators grow much more quickly and therefore must eat more. Furthermore, although the predators eat much less when mature, the wastage of growing individuals is large and this means that the age distribution of predator populations is loaded in favour of the most voracious age groups.

The end result of this admittedly complex calculation is highly significant, and not all in favour of the plaice. It turns out that the plaice eat only about 6–7 per cent of the total amount of available fish food, while the lower predators account for the rest. In terms of fishery economics this is lamentable, but now that we know about it we can take action in a rational manner. The first corollary to these figures is that in most cases we must not attempt to fertilise landlocked waters, because the undesirable animals will profit more from increased primary production than will the commercially valuable fish. We must look elsewhere for a means of effective action. An example that shows how difficult this is may be seen in an attempt to improve production in shallow waters off the Atlantic coast of the United States and Canada, where the density of sea stars was up to fifteen specimens per square metre. Special boats were constructed to suck up the entire bottom population through a sort of gigantic vacuum cleaner and to sort out the sea stars. But this is not much use when a single medium-sized sea star can produce two and a half million eggs which, as pelagic larvae, can later swarm over and land like paratroopers on the very areas just cleared.

We can take some comfort in considering what can be done in shallow water where it is possible to control what is happening. There are shallow lagoons, some of them even brackish, as far apart as Indonesia and Arcachon in France, which can produce from 11 to 50 tonnes of live fish per square kilometre per annum. Brackish ponds in Formosa do even better with 90 to 135 tonnes. This spectacular production is achieved by exploiting a remarkable fresh-water fish called *Tilapia,* which can also live in brackish water. In many ways this is a 'wonder fish' in that it leaves other fish in peace; it can be acclimatised to survive at temperatures up to 38°C (equivalent to a hot bath), and although it grows best on an animal diet, it lives mainly on algae, detritus and plankton, and is thus closer to the bottom of the food-chain than almost any other fish. *Tilapia* can reach sexual maturity in about three months, and in extremely favourable conditions it can grow to a weight of 850 grams in a year. Recent Russian experiments on the productivity of two pairs of *Tilapia mossambica* (perhaps the most prolific species) showed that even when less than one-quarter of their offspring were under cultivation, the initial weight of the two pairs increased by a factor of 224 in eight months.

10·7 Top Weekly consumption of soft-shell clams (*Mya*) by the drilling, predatory snail *Polinices* (=*Natica*) *duplicatus* off New England. (Poul H. Winther.)

10·8 Bottom Weekly consumption by a newly hatched oyster drill (*Urosalpinx*) when feeding on oyster spat (*Ostrea edulis*) in southern England. (Poul H. Winther.)

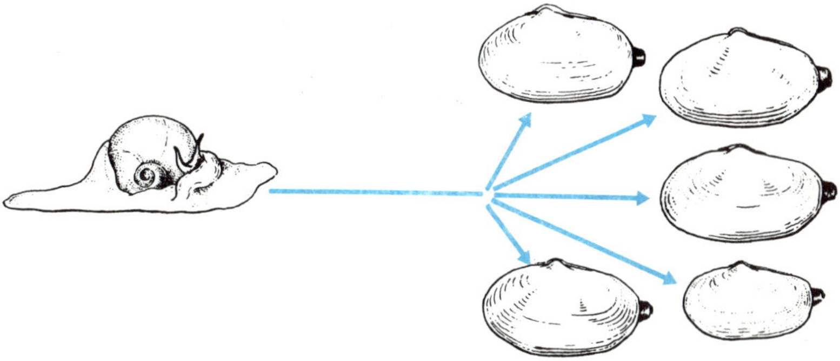

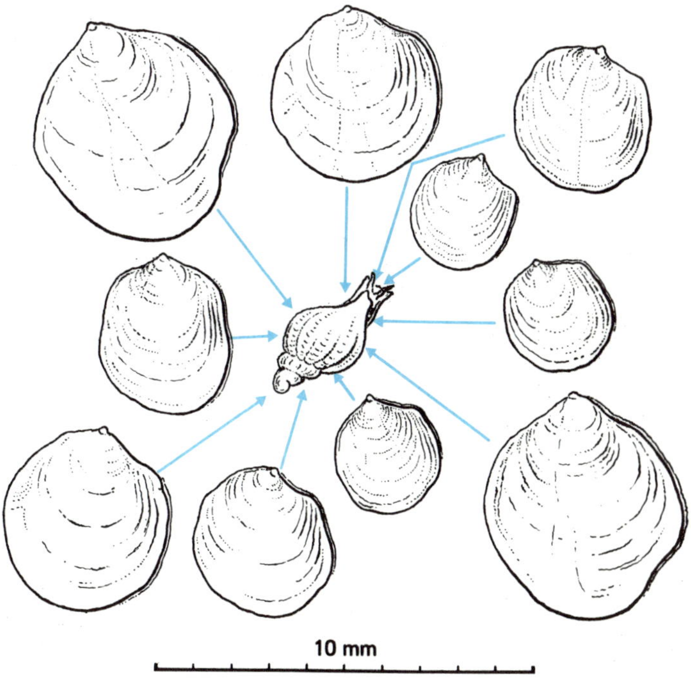

Tilapia breeds so prodigiously that even in a well-run breeding pond it is almost impossible to control, overcrowding ensues, the individual fish grow too slowly, and the profit drops. Nevertheless breeding in ponds and shallow sea water can be profitable to the extent that sheltered and shallow water areas can yield more protein per unit area than a field used for raising cattle.

Though *Tilapia* heads the list for productivity, there are others that can be 'farmed' in the same way. The fish *Pogonias cromis*, cultivated in salt-water ponds in South Carolina, has shown that it can multiply 21-fold in nine months. Fish of the mullet family (*Mugil*) also appear to be promising; they are fast growers, can stand changes in salinity, and like *Tilapia* can eat algae and detritus.

These successes in shallow-water fish farming point to a conclusion; namely that it is a great mistake to reclaim shallow-water coastal areas in order to convert them into arable land, or even into pasture for raising cattle. It has been shown that a well cultivated brackish water area can produce three times as much protein as the same area of farmland.

Fishery regulations and control We can perhaps argue that we have not entirely neglected our fishery regulations. We have laid down minimum sizes for the fish that may be caught, so that the population of edible fish can propagate before going into the net. But even in

10·9 The weekly consumption of a shore crab (*Carcinides maenas*) when feeding off soft-shelled clams (*Mya*) off the Atlantic coast of Canada (Poul H. Winther.)

framing such regulations, fishery biologists run into problems that are far more complicated than one imagines. It is generally agreed, for instance, that fish must be partially or completely left alone while they are spawning; the more females that spawn successfully, the greater the number of developing young and hence the greater the potential edible fish catch in the future. This seems obvious, but it is not necessarily so, as witness a Californian experience. In the course of an intensive hunt for squids, doubts were raised as to whether the squids were perhaps being over-exploited. During the breeding season, when these animals gather on the banks in huge numbers, frogmen were sent down to see what was happening. They found that the eggs had been laid in such thick layers that only the topmost tenth received enough oxygen for normal development. The lower layers were already bluish-black, smelt of hydrogen sulphide and were obviously lost. In effect, too many squids were laying too many eggs, and it was therefore possible to catch many more squids without endangering the stock.

The same thing is also true of fish. We have seen that each female normally spawns a very large number of eggs, so that under favourable conditions the population can be maintained by surprisingly few females. As a rule the amount of protection afforded by man is trifling compared with the impact of natural conditions, such as predators, or quite small changes of temperature. This is not to say that protection is worthless; it means rather that we should consider carefully for each species just how protection should be applied, at what times of the year and for how long.

In some cases, protection is of great value. For example, the Dogger Bank in the North Sea is a valuable fishing ground for plaice. During the second world war fishing became almost impossible, and when the war was over it was found that the plaice population had markedly increased. But it was also found that with the resumption of fishing the population soon reverted to its pre-war density, and today in certain areas of the North Sea 70 per cent of tagged plaice are being recovered inside twelve months. This indicates that the area cannot support any more intensive fishing than is currently taking place. But in considering what form fishery protection should take we have to bear certain factors in mind. One of these is that a food-fish population should be thinned out in the same way that trees are felled in a well managed forest. In particular it is desirable to eliminate *old* fish, which grow slowly, utilise food inefficiently, and may even eat young members of their own species. Take for example the case of the American sea fish *Morone,* of the perch family. It has been calculated that by the time this fish is twenty years old it is utilising only 1 per cent of its food for further growth, whereas at the age of four years it was converting 25 per cent of its food intake into weight increase. Now it is probable that the twenty-year-old female spawns over two million eggs while the younger fish produces only half a million. But the latter figure is quite enough to maintain the population, and it follows that the existence of the old fish is a luxury that a well-managed fishery cannot afford.

Regarding fishing control in general, it is important to understand precisely what so-called overfishing means. Usually it is the result of too many fishermen hunting too few fish. But it does not mean that the fish of a particular species are exterminated, nor indeed are they in any such danger. The law of supply and demand operates very effectively in fisheries; when a fish population is depleted the result is simply that fishing becomes uneconomic. Ideally the number of fish caught should exactly balance the number produced. A healthy industry should operate at this level, but as so often happens, particularly when a new fishing area is opened up, it is besieged by too many fishermen and the catch becomes less and less. More significantly, the catch *per boat* diminishes. The Canadian fish biologist, J. L. Hart, gave a good example of this in connection with the highly competitive British Columbia salmon fishery, when in

1950 he pointed out that a smaller number of fishermen could bring in the same number of fish. At the same time we must make a sharp distinction between overfishing and extermination. We know of several cases on land where ruthless exploitation of certain animals has led to total extinction or to the last-minute rescue of a species by strict game preservation laws. But one need never fear that the same thing could happen in the sea; investment in modern fishing boats and gear is so expensive that a fishery will run at a deficit long before the fish themselves are exterminated. This same argument also applies to whaling, which has been more ruthlessly pursued than any other kind of 'fishing'; one can be sure that the soaring cost per whale caught will be more effective in preserving the surviving animals than any regulations imposed by the International Whaling Commission.

Unplanned impacts

At this stage we should stress that human activity unaccompanied by scientific awareness of the consequences can produce unexpected situations which are quite beyond our control. A striking example is to be seen in the long-term consequences of digging the Suez Canal.

When the canal was opened in 1869 it provided a direct link between the tropical Red Sea and the subtropical eastern Mediterranean; because of a slight difference in levels of the two seas, current flowed (and still flows) in a northerly direction. It would be expected therefore that there would be a migration of Indo-Pacific species into the Mediterranean. However, a biological survey made in 1929, sixty years later, showed that only fifteen species had migrated northward, while none had moved in the opposite direction. There were in fact two major obstacles to migration. The first was that part of the 162 kilometre length of the canal consists of the

Great Bitter Lake, at the bottom of which was a deposit of 970 million tons of salt. The slow current dissolved this salt and gave the passing canal water a salinity of between 76–80‰, too much even for Red Sea animals.

Even if a Red Sea migrant survived this salinity barrier, there was yet another barrier, in the shape of a sharp reduction in salinity in the region of Port Said, caused by the waters of the Nile. Here the salinity was reduced to as little as 26‰ during the Nile floods (and was still much reduced at other times of the year): as we have seen in reference to the Baltic, very few truly marine animals can tolerate brackish conditions. In short, there existed a near-perfect biological barrier – at least that appeared to be the situation in 1929. Since then, however, several things have happened. The canal has been progressively deepened and widened, so that by 1963 the flow had increased by 50 per cent. The mass of salt in the Great Bitter Lake had largely dissolved and the salinity has been reduced to 43–46‰, close to that of the Red Sea at 42–43‰. And the Aswan High Dam now retains so much fresh water from the Nile that the salinity at Port Said has risen to 39 or even 40‰.

What are the consequences of these massive changes? Not all the alterations of environment have happened simultaneously (the reduction in Nile water is quite recent), but over the last forty years the trend is already clear to see. About nine times more species have passed through the canal in the last forty years than did in the previous sixty years. It should be mentioned that apart from the removal of the salinity barriers, the increased current velocity may soon be enabling larval forms to make the journey, whereas formerly the migrations were probably carried out only by adults. However that may be, the results in economic terms are already promising. Many Indo-Pacific animals have established themselves in the south-eastern Mediterranean, and some are now so plentiful that they are fished on a commercial scale. These include a species of sprat

(*Dussumieria*), a barracuda (*Sphyraena*), a crab (*Neptunus*), a prawn (*Penaeus*) and the pearl oyster (*Pinctada*). So far the results of the free connection between the two seas seem to have been wholly favourable, but such is the complexity of marine ecology that there is always the possibility that some unwanted animal will penetrate the Mediterranean and upset the present balance of nature. Indeed, one of the strongest arguments that have been put forward against the building of a sea-level canal across the Isthmus of Panama is that there is a danger of Pacific invaders upsetting the ecology of the entire Caribbean.

Changes in the Black Sea As we have just seen, a 'faunal revolution' is taking place in the Mediterranean; but we can expect a similar upheaval in the Black Sea. At the moment the Black Sea is heavily stratified, owing to the inflow of fresh water from Russian rivers (as well as the Danube): there is little or no mixing between the oxygenated upper layer and the sterile, anoxic and more saline waters below 200 metres. However, all this will change in the late 1970s, when Russia's combined hydro-electric and irrigation schemes will have reduced the fresh-water flow by approximately 150 cubic kilometres per annum – an amount comparable to the inflow of saline water through the Bosphorus at the present time. The reduction in freshwater flow will increase the salinity of the Black Sea (at present averaging only 18‰) thus enabling vertical circulation to oxygenate the bottom water, and improving conditions for animal life at greater depths. In the long term the Black Sea should be prepared to receive migrants from the Red Sea.

Ecological chain reactions We have already sounded a warning that it is not only beneficial animals but also noxious ones that may migrate during a translocation of fauna. An interesting example is to be found in the Black Sea, where a large Japanese predatory snail, *Rapana thomasiana,* was accidentally introduced in 1947 into the Novorossisk area, on the north-east coast of the Soviet Union. Just how it arrived is something of a mystery, but it seems likely that its tough and resistant egg capsules made the journey adhering to the shells of oysters, also imported from Japan. Today *Rapana* is widely distributed over a great many coastal areas, and since its favourite food consists of bivalves, many oyster and mussel beds have been totally destroyed. And of course the next consequence was a reduction in bottom-feeding fish which lived mainly on just these bivalves.

But the story does not end here. The next event was a massive invasion of a sand eel, *Ammodytes cicerellus,* which moved in through the Bosphorus from the Mediterranean, and adopted the pelagic larvae of *Rapana* as its main diet. Next, the fish *Sargus,* which had hitherto fed on worms and crustaceans, changed its diet to *Ammodytes.* And so it goes on; we cannot yet tell what the final balance of power initiated by the Japanese mollusc will be. There is, however, one amusing side-effect: the large Mediterranean hermit crab *Clibanarius* was formerly represented in the Black Sea as a 'dwarf' race, the reason being that only small snail shells were available in which it could hide its soft abdominal parts. Now, however, there is a good supply of empty *Rapana* shells, the hermit crab can expand and attain the same impressive size as its Mediterranean relative.

Improving our exploitation of the sea

Most of our control over farming marine life concerns only the shallow water areas lying just off our coasts. The immense areas of the large oceans can never be cultivated; they will remain wild areas, and only as hunters can we exploit them. This brings us back to the same problem: in order to utilise fish in a rational way we must get to know their habits much better than we do today. For example, let us consider a fish's sense of colour. This can influence the catch to a large degree. We know that shoals of the fish *Jenkinsia* will not approach a dark object as closely as a light one, and one must accordingly choose the right colour of bait or net. As a general rule it has been found that green nets give the best yield in phytoplankton-rich water, and that blue nets work best in phytoplankton-poor water. We have seen how much the schooling of fish depends on light, but most fishes react only to a very narrow part of the spectrum which is 'optimal' for them. Thus the colour of the light is of great importance. Mediterranean fishermen have for a long time attracted sardine schools by artificial light; they know very well that white light is more effective than yellow light, and that the fish are much more stimulated by artificial light on moonless nights than at full moon. Promising results in attracting schools of certain species of fish and squids have also recently been obtained by lowering fluorescent lamps into the sea.

In the case of other fish it is quite certain that they are able to sense a temperature change of as little as 0·003°C and a salinity difference of 0·2‰, thus rivalling the most sensitive hydrographic instruments yet devised by man. And knowing that they make use of these abilities, we can utilise the reactions of the fish in order to track them down. In the Lofoten fisheries off Norway, temperature measurements of sea water play a most important part, and one can calculate almost to the day where the fish will be and exactly when they will enter the fishing grounds. Thus the whole fishing fleet can be notified and ready to strike at the right moment. We also know that a single species of fish can react in different ways at different temperatures; for instance, it may escape below the net in cold water or use the escape route above the net in warmer water.

Let us take a look at something that might seem to be purely academic – the ability of fish to transmit sound. Fish noises suddenly became an important field of investigation during the second world war, especially when it was discovered that certain rather common and 'harmless' fish along the coasts of the United States could produce noises loud enough to explode acoustic mines. Hitherto we had been only vaguely aware of these noises, as the intensity of sound in water is diminished by a factor of 1,000 when it passes from water to air. Today fish sounds are recorded on tape with the aid of underwater microphones, and an entirely new interpretation of their meaning has been obtained. We now know that out of sixty species of fish investigated off New England, fifty-six produce sound, and that such widely different animals as sea urchins, squids, shrimps, crabs, fishes, sea lions and porpoises all contribute to the underwater noise.

The underwater explorer Hans Hass was able to show in a film how a certain fish species will gather in a circle of well defined radius around a submerged loudspeaker playing a Strauss waltz. Most fishermen have hardly given such behaviour patterns a thought, but all the same it is not impossible that within a few years we may create new and profitable fishing methods by attracting fish shoals with recordings of their own sounds. We may also learn to interpret the different kinds of sounds to locate a fish school and to find out *what* species the school consists of by listening to the sounds emitted. Perhaps, too, we may even find sounds which we can broadcast under water and which may frighten porpoises and dolphins away from fish traps and nets, where they do much damage.

Finally, during recent years there have been demands to exploit the reactions of fish to electricity. The method of stunning fresh-water fish by electric shock has proved very effective, and in many small streams the number of fish taken has been much greater than one would have imagined. In the sea, however, prospects are somewhat less promising. The sea has a much larger conductivity than fresh water, and there is a great difference between sending an electric current through a comparatively small water mass in a fresh-water stream, and sending one through the boundless areas of the sea. Even if it were not too expensive it would probably be more dangerous to the operator than to the fish. Quite recently, however, some interesting possibilities have come to light, in connection with electricity being used with trawling gear. Experiments at George's Bank off Massachusetts have shown that the catch of cod and haddock was unaffected but that with certain other species, such as whiting, the yield increased four- to eightfold.

When it comes to fishing for shrimps and prawns, electric-shock techniques may prove to be little short of revolutionary. The large tropical shallow-water species usually dig down into the sea bed by day, to reappear on the bottom surface at dusk, so that fishing has to be done at night. Also, during the cold season, when the temperature even in these waters is below average, the shrimps dig themselves in at night too and are very sluggish in their movements. Experiments in the laboratory and at sea have now shown that a shrimp-trawl fitted with electrodes which scrape the bottom and transmit four to five pulses per second of a three volt direct current into the substrate galvanise shrimps into leaping seven to eight centimetres above the bottom. This is enough to get them into the net, and comparisons between trawls with and without electrodes have shown that in broad daylight the former can catch five to twelve times more than the latter. This means that shrimp-fishing in future can take place at all seasons, by day as well as by night; and if a moderately priced commercial version of the system could be developed, it would at least double the efficiency of the shrimp fleet. Similar responses to electric current have been found for crabs of the genus *Portunus*.

Another method of fishing which shows promise exploits the fact that many fish, and especially herring, are frightened to pass through a curtain of air bubbles. We can explain the technique most easily by referring to a method of catching herring which is practised off the coast of Maine in the United States. Here the fishing has been done by permanently attached ground nets or by long nets hanging from the surface. It was possible to isolate narrow, shallow-water bays and fjords by this method. But often the herring escaped because they were moving too deep for the twenty metre net to intercept them. The cost of extending the working depth of the net was uneconomical but an artificial extension could be made by laying out a perforated plastic pipe and blowing compressed air so as to make a curtain of bubbles. Such a pipe can be used in a number of ways, but a simple and ingenious method that originated in 1957 consisted of laying out 700 metres of pipe in such a way as to force a shoal of herring to swim along the bubble curtain until they reached shallow water and were trapped by the ground net. Other possibilities with bubble curtains include manoeuvring the pipe so as to encircle a shoal in deep water and then drawing the entire shoal into shallow water where existing nets will effectively catch the herrings.

Finally, in this review of improved fishing possibilities, it is possible to combine several new methods at once. For example, schools of fish may be attracted by using the right kind of light, by emitting the correct 'signature tune' of a species through submerged loudspeakers (perhaps also by training porpoises or sea lions to take on the function of sheep-dogs) by encircling the shoal, immobilising it with electric shocks, and finally sucking the fish up into the factory ship through giant funnels. Such systems are already being employed by Russian and Japanese fishery experts with great success.

Euphausiids as a food source? Earlier in this book we referred to copepods and euphausiids as the outstanding converters of phytoplankton into protein in the deep oceans. We have seen that copepods are the main fish food in the northern hemisphere, but what of the euphausiids in the Antarctic? We know that the southern baleen whales live almost exclusively on these oceanic prawns, simply by cruising slowly, with the mouth slightly open and filtering euphausiids. Perhaps the less said about southern whales the better, since the southern blue whale is greatly reduced in numbers through over exploitation, and the smaller and less profitable baleen whales – fin, seit and humpback – are now also becoming gravely depleted. But the fact remains that until recently their consumption of euphausiids must have been enormous. To take an example: in 1950 the capture of baleen whales was equivalent in meat value and calorific content to 1,800,000 average sized bulls. Such a figure postulates a fantastic supply of (mainly) *Euphausia superba*.

It is tempting to consider the possibility of imitating the baleen whale and fishing for euphausiids as a direct food source. Certainly their size (that of a small shrimp) makes this a far more practicable proposition than the idea of direct fishing for copepods, because a larger-meshed plankton net could be used which would be less likely to become clogged. The idea of direct plankton fishing is even more attractive when one considers that recent calculations indicate that a hundred million tonnes of euphausiids per year could be harvested from the Antarctic alone. Even so, it is still doubtful whether direct plankton fishing would be economic. It is one thing to let whales filter euphausiids for several years and then to 'cash in' on their efforts with one well-aimed shot of a harpoon; it is quite another to operate a fully manned ship with huge plankton nets and expect to gather in protein at an economic price.

One other possibility would be to introduce into the Antarctic some species of northern fish which could absorb the existing surplus of plankton. If it proved possible to do this, there could be a saving of millions of tonnes per annum of protein food for mankind. But it must be said that even if such a northern fish were to 'catch on' and flourish, the catching of these fish would present very great problems.

The Antarctic is a very large oceanic area and most of the waters are deep. There would be no concentration of fish in relatively shallow waters except round South Georgia and a few other islands and it would probably be necessary to develop some technique of intensive deep-sea fishing of a kind for which there is no existing parallel.

Food in the scattering layer

We have already mentioned the scattering layer, which once confused scientists exploring the sea with echo-sounding apparatus. At one time there appeared to be a good possibility of exploiting this layer, where a great deal of marine life appears to be concentrated all over the world in a relatively narrow stratum. In the past ten years, however, we have become more sceptical, since the animals of this layer do not have to be very close together to produce an effective echo. Nevertheless, marine scientists are agreed that one of the fish in the layer, *Cyclothone*, is the most numerous species on earth; and it is not fished at all. Failing some method of catching this fish in economic quantities we can fall back on the self-evident fact that this multitude of fish must consume at least ten times their own weight of copepods and euphausiids. If we could learn how to fish the euphausiids at night when they migrate towards the surface, we might be able to make this source of food very profitable. We are still faced with the economics of plankton fishing which, at present, do not look very hopeful. But we cannot neglect any possibility of trying to alleviate the world protein shortage by intensifying our attention on any and every possibility of exploiting the life of the sea for man's benefit.

11 Some areas of ignorance

It may seem that we already know a great deal about life in the sea, and in a sense this is true. But what we know is only a small fraction of what remains to be learned, and if marine biologists are to play their part in increasing food supplies, they must make efforts on a hitherto unknown scale to learn the biology of sea animals.

Parasites

Let us take as one example our ignorance of parasites. We know, for example that the plaice is known to harbour fifty-three known multicellular parasites. The cod has seventy, the sockeye salmon has forty-three and the herring of the Black Sea alone has thirty-two. These four species of fish are edible and commercially important, and this is why we know so much about them. But they are the only fishes out of 20,000 known species that have been thoroughly investigated.

It can safely be assumed that all marine creatures suffer from parasites. For instance the 2 millimetre long mud snail *Hydrobia,* common along Scandinavian coasts, has at least eleven species. More than 100 species of parasitic copepods infest molluscs, and more than 117 genera of molluscs are known to be infected with them. Furthermore, copepods themselves also have parasites, and as mentioned before, the commonest copepod *Calanus finmarchicus* carries at least twelve. From these few specimen figures we can calculate that the number of parasites is much greater than the number of host animals. It goes without saying that in general parasites are detrimental to the growth, well-being and propagation abilities of their hosts.

Many species of fish, particularly in warm waters are seriously bothered by parasites, and in order to survive it is essential that they get rid of them. This applies particularly to *ectoparasites,* organisms feeding on the skin, the gill rakers and mouth parts. It is

interesting to find that forty-five species of fish and six species of shrimps have specialised as 'cleaners', either part-time or in some cases getting all their food in this way. The cleaner fish are usually gaily coloured to attract their customers, and they wait about for infested fish to come to them regularly for a thorough cleaning.

Diseases of fish

We have been talking about multicellular parasites in order to distinguish them from the protozoa, bacteria and viruses which also affect marine animals. As regards the viruses, we know almost nothing about them and hence we have no idea how to combat them. Today we should be as helpless in fighting an epidemic among fishes as were the physicians of the Middle Ages in tackling the plague. Imagine the result if an epidemic (similar to that which attacked eel grass in the 1930s) were to occur among the whole cod population of the North Atlantic. This is by no means impossible, and we would be quite helpless in the face of a disaster that might wipe out 25 per cent of the world's fishery.

Animals dangerous to man

A good deal of research has recently gone into studying dangerous animals such as shark and barracuda, and various means have been devised for protecting humans in shark-infested waters. But we have

11·1 Many fish are heavily infested with parasites, as with this ocean perch (*Sebastes marinus*) with its large parasite copepods (*Sphyrion lumpi*). (After E. S. Iversen 1964.)

almost everything to learn about *poisonous* marine animals, poisonous that is either to touch or to eat. We have mentioned the stinging cells of cubomedusae and siphonophores which can kill a human being, the burning sensation imparted by stinging corals, the poisonous bite of octopuses, and the venomous sting of the crown-of-thorns sea star; we have also talked of the 'red tide' dinoflagellate (*Gonyaulax*), which causes the mass death of fish and fish-eating birds, as well as contaminating molluscs so that they are dangerous for man to eat. There are many other such creatures, for instance some species of the tropical snail, *Conus*, which have a bite that can kill a man; or the colonies of a North Sea polyzoan *Alcyonidium* which can generate an eczema (called Dogger Bank itch) which sometimes compels fishermen to give up their trade. Add to these the occupational diseases of coral fishers in the Mediterranean and of sponge divers in the tropics, the dermatitis caused by huge swarms of trematode larvae, and the irritating effect of accidentally touching certain polychaete worms, and even then we have hardly begun to list the poisonous fish. In fact the tropics contain over 3,000 species of poisonous fish – poisonous to touch or to eat. The moray eel which lurks in rock crevices and in coral has a poisonous bite; sting rays have poison spikes in their tails; six families of the multi-coloured fish of coral reefs have poison spines, some of them fatal to man; and there are many fish which are poisonous to eat – the balloon fish was only one example.

Healing substances from the sea

After this recital of a few of the noxious elements in marine life, it is good to be able to report that there are indications in the most recent publications that the sea is a rich source of medically useful compounds. The noxious *Gonyaulax*, for example, is now found to contain three different antibiotics; certain unicellular green algae

11·2 The greater weever *Trachinus draco*, with grooved poison spines on the first dorsal fin and on its operculum, can cause extremely painful stings and is feared by North Atlantic fishermen. (After R. Bott 1939.)

contain a substance called halosphaerin which destroys bacteria of the staphylococcus type, as does also an antibiotic called ectyonin from the red-beard sponge, which can even kill penicillin-resistant tuberculosis bacilli.

Molluscs such as abalones and oysters contain a drug called paolin II which appears to be effective against viruses, while holothurin from sea cucumbers has shown promising results against certain types of cancer. A toxic substance from the weever fish reduces the heart beat (like digitalis from foxgloves), another toxin from the stone fish reduces blood pressure, and tetrodotoxin from the balloon fish is the best pain killer that we know of so far.

Shortage of manpower

The discovery that antibiotics can be derived from marine animals points directly to the need for further research and a whole new branch of marine biology is being born. But who is going to do the work? Similarly, the experimental breeding of micro-algae as a source of food shows promise, but there has been only moderate success so far (and here we should remind ourselves that out of ten thousand known species of algae less than one hundred have been the subject of experimental work). In addition to basic marine biological research, new problems are constantly arising; some of them grow out of new discoveries, others are the result of developments in other fields, such as the disposal of radio-active waste, the ever-increasing need to protect marine life from the devastating effects of accidental pollution of the sea with oil from tankers, and from the cumulative effects of pesticides.

It should by now be obvious that in order to develop and conserve the resources of the sea, the world needs an immense increase in the numbers of marine biologists, not to mention more accommodation and equipment both on shore and at sea. This need seems to have been appreciated in Russia and the United States, where large sums of money have been appropriated for training and research. But money and facilities are not enough in themselves. It takes at least ten years of university and postgraduate training to produce a competent marine biologist or oceanographer, and the demand is rising so rapidly that even with the latest training programmes we shall have far too few marine scientists for many years to come. It is on the existing training programmes that the whole future of oceanic research depends. The new generation of marine biologists faces an ever-growing number of large and pressing problems, the solutions to which are of importance to the whole of humanity.

Bibliography

Anderson, H.T., *The Biology of Marine Mammals*, London and New York, 1969.

Barme, M., 'Venomous Sea Snakes of Vietnam and their Venoms', in Keegan and Macfarlane, *Venomous and Poisonous Animals and Noxious Plants of the Pacific Region*, London and New York, 1963.

Barnes, H., *Oceanography and Marine Biology: a Book of Techniques*, London and New York, 1959.

Borgstrom, G., *Fish as Food*, London and New York, 1961.

Bruun, A. (*et al.*), *The 'Galathea' Deep Sea Expedition 1950–2*, London and New York, 1956.

Buzzatti-Traverso, A.A. (ed.), *Perspectives in Marine Biology*, Los Angeles, 1968.

Carrington, R., *A Biography of the Sea*, London and New York, 1960.

Carson, R.L., *The Sea Around Us*, London and New York, 1951.

Carson, R.L., *The Edge of the Sea*, London and New York, 1955.

Clarke, G.L., *Elements of Ecology*, London and New York, 1965.

Cousteau, J.-Y. and Dumas, F., *The Silent World*, London and New York, 1952.

Darnell, R.M., 'The Organic Detritus Problem', in Lauff, G.H. (ed.), *Estuaries*, London and New York, 1967.

Darwin, Charles, *The Voyage of a Naturalist round the World in HMS 'Beagle'*, London and New York, 1839.

Eales, N.B., *The Littoral Fauna of Great Britain*, Cambridge and New York, 1939.

Ekman, S., *Zoogeography of the Sea*, London and New York, 1953.

Flattely, F.W. and Walton, C.L., *The Biology of the Sea-Shore*, London and New York, 1946.

Fraser, J., *Nature Adrift. The Story of Marine Plankton*, London, 1962.

Graham, M. (ed.), *Sea Fisheries*, London and New York, 1956.

Green, J., *The Biology of Estuarine Animals*, London and New York, 1968.

Halstead, B.W., *Dangerous Marine Animals*, Cambridge (Maryland), 1959.

Hardy, A.C., *The Open Sea* I: *The World of Plankton*, London and New York, 1956.

Hardy, A.C., *The Open Sea* II: *Fish and Fisheries*, London and New York, 1959.

Hardy, A. C., *Great Waters*, London and New York, 1967.

Harvey, H. W., *The Chemistry and Fertility of Sea Waters*, Cambridge and New York, 1955.

Hedgpeth, Joel W. (ed.), *Treatise on Marine Ecology and Paleoecology* 1: *Ecology* (Mem. Geol. Soc. Amer., 67 (1)), 1957.

Heen, E. and Krenzer, R. (eds), *Fish in Nutrition*, London, 1962.

Hill, M. N. (ed.), *The Sea* (3 vols), London and New York, 1962-3.

Idyll, C. P., *Abyss, the Deep-Sea and the Creatures that Live in it*, New York, 1964.

Jørgensen, C. Barker, *Biology of Suspension Feeding*, London and New York, 1966.

Lewis, J. R., *The Ecology of Rocky Shores*, London, 1964.

MacGinitie, G. and N., *Natural History of Marine Animals*, London and New York, 1949.

Marshall, N. B., *Aspects of Deep Sea Biology*, London and New York, 1958.

Marshall, N. B., *The Life of Fishes*, London and New York, 1965.

Marshall, S. and Orr, A. P., *The Biology of a Marine Copepod*, London, 1955.

Mauchline, J. and Fisher, L. R., 'The Biology of Euphausiids', in Russell, F. S. and Yonge, C. M. (eds), *Advances in Marine Biology* (7), 1969.

Menzies, R. J., 'Conditions for the Existence of Life on the Abyssal Sea Floor', in Barnes, H. (ed.), *Oceanography and Marine Biology* (3), (1965).

Miner, R. W., *Field Book of Sea Shore Life*, London and New York, 1950.

Moore, H. B., *Marine Ecology*, London and New York, 1958.

Murray, J. and Hjort, J., *The Depths of the Ocean*, London, 1912.

Newell, G. E. and R. C., *Marine Plankton: A Practical Guide*, London and New York, 1963.

Newell, R. C., *Biology of Intertidal Animals*, London, 1970.

Nicol, J. A. C., *The Biology of Marine Animals*, London and New York, 1960.

Nikolsky, G. V., *The Ecology of Fishes*, London and New York, 1963.

Norris, K. S. (ed.), *Whales, Dolphins, and Porpoises*, Los Angeles, 1966.

Odum, E. P., *Fundamentals of Ecology* (2nd ed.), London and Philadelphia, 1967.

Ommanney, F. D., *The Ocean*, London and New York, 1949.

Petersen, C. G. Joh., *The Sea Bottom and its Production of Fishfood*, Copenhagen, 1918.

Piccard, Jacques and Dietz, R.S., *Seven Miles Down*, London and New York, 1962.

Ray, D.L., *Marine Boring and Fouling Organisms*, Seattle, 1959.

Raymont, J.E.G., *Plankton and Productivity in the Oceans*, London and New York, 1963.

Riedl, R., *Fauna und Flora der Adria*, Hamburg, 1963.

Riedl, R., *Biologie der Meereshöhlen. Topographie, Faunistik und Ökologie eines unterseeischen Lebensraumes*, Hamburg, 1966.

Russell, F.S. and Yonge, C.M., *The Seas*, London and New York, 1949.

Russell, R.C.H. and Macmillan, D.H., *Waves and Tides*, London and New York, 1952.

Sverdrup, H.W., Johnson, M.W. and Fleming, R.H., *The Oceans, their Physics, Chemistry, and General Biology*, London and New York, 1946.

Thorson, G., 'Length of Pelagic Larval Life in Marine Bottom Invertebrates as Related to Larval Transport by Ocean Currents', in Sears, Mary (ed.), *Oceanography*, 1961.

Walford, L.A., *Living Resources of the Sea*, New York, 1958.

Wickstead, J.H., *An Introduction to the Study of Tropical Plankton*, London and New York, 1965.

Wilson, D.P., *Life of the Shore and Shallow Sea*, London, 1951.

Wimpenny, R.S., *The Plankton of the Seas*, London and New York, 1966.

Yonge, C.M., *The Sea Shore*, London and New York, 1949.

Yonge, C.M., 'The Biology of Coral Reefs', in Russell, F.S. (ed.), *Advances in Marine Biology* (1), 1963.

Zenkevitch, L., *Biology of the Seas of the USSR*, London and New York, 1963.

In addition, the following journals are recommended to the interested reader: *Ecology, New Scientist, Journal of the Marine Biological Association, Journal of Experimental Biology, Science, Nature, Sea Frontiers, Fishing News International, Oceanus, Ophelia, Scientific American, Journal of Animal Ecology, Biological Revue, Copeia, Natural History.*

Acknowledgments

My husband, Professor Gunnar Thorson, died on 25 January 1971. I know that he felt most grateful to his colleagues and their publishers for their permission to use illustrations from their publications, though in fact it has not been possible for me to check whether written permission had been granted in every case. On his behalf, I would like to thank the following authors and their publishers: J.C.C.Nicol, for figure 1·2; D.E.McAllister, 1·3; T.J.Smayda, 1·4; J.Fraser, 2·2, 5·5; D.Wilson, 2·5; W.M.Stephens, 2·6, 8·7; E.Thörner and W.E.Ankel, 2·7; Kai Olsen, 2·8, 5·1, 5·2, 5·5, 7·9, 8·10; Logan Smith, 3·1; G.S.Carter, 3·2; H.H.Hildebrand, 3·3; A.Cerruti, 4·5; F.S.Russell, 5·3; E.C.Pope and P.M.McDonald, 6·1, 7·16, 7·17; B.A.Wooding, 7·1; C. Hauenschild, A.Fisher and D.K.Hofmann, 7·2; H.Battle, 7·3; B.A.Wade, 7·4; J.R.Lewis, 7·6, 8·3; C.M.Yonge, 7·10, 7·11; T.F.Goreau, 7·12; H.W. Fricke, 7·13; A.S.Pearse, 7·14, 7·15; A.Seilacher, 7·18; E.E.Boschi, 7·19; V.A.Harris, 7·21; J.H.McLean, 8·4; S.M.Lodge, 8·5; H.Kühl, 8·9; E. Zeigelmeier, 8·14; P.H.Winther, 8·15, 10·7, 10·8, 10·9; A.Margolin, 8·16; C.G.Joh.Petersen, 8·17, 8·20, 10·3; A.D.Ansell, 10·1; L.A.Zenkewitsch, 10·2; H.Feder, 10·4; E.S.Iversen, 11·1; R.Bott, 11·2.

I would also like to thank the photographer Mr Holger Knudsen, and the artist Mr Kai Olsen, both at the Helsingør Marine Laboratory, for their great help in illustrating this book, which is my late husband's final contribution to marine biology.

Ellen Thorson
Helsingør, 3 March 1971

Index